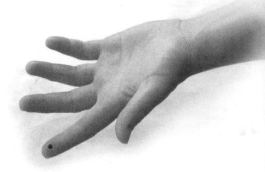

Personal Genomics
and
Personalized Medicine

To my parents

Contents

CHAPTER 1

Introduction

As he admired his newborn daughter, Hugh Rienhoff had an unwelcome sense of déjà vu. Trained as a clinical geneticist, he immediately noticed that Beatrice had the elongated feet and long fingers characteristic of a rare genetic disease known as Marfan syndrome. Over time "Bea" exhibited a marked lack of musculature and weight gain, but she did not have some of the life threatening aspects of that syndrome. The related Beals and Loeys-Dietz syndromes were similarly ruled out. Five years and many consultations later, Bea's condition remains undiagnosed.

Although her collection of signs and symptoms are very rare, Beatrice's situation is not. Every year thousands of children are born with diseases that defy classification and treatment. Such children typically have to "make do" with drugs and treatments that only address their condition partially or palliatively. Some receive the wrong treatment. Many live short, difficult lives.

Beatrice is lucky. Her father is a highly educated specialist in genetic disorders with extensive connections in the medical community. After reading the relevant research literature and discussing his daughter's case with many of the leading specialists, Rienhoff noted that all three of the above genetic disorders are thought to involve malfunctions of a signaling pathway for a family of hormones known as Transforming Growth Factor β (TGF-β). Could it be that the symptoms Bea shares with these disorders are also due to dysregulated TGF-β signaling? This is a testable hypothesis offering a potential diagnosis; and diagnosis is a precondition to management and treatment.

At the time of writing this book, Rienhoff is sequencing many genes in himself, his wife and his daughter to look for mutations that may affect the TGF-β pathway (directly or indirectly).[1] Scientific research is always a process of trial and error. Rienhoff's current

---

[1] A full account of Hugh and Beatrice Rienhoff's odyssey appears in a January 2009 article in *Wired Magazine*, available at http://tinyurl.com/beatriceDNA.

hypothesis may or may not turn out to be correct. If it fails, the symptoms-to-pathways-to-genes methodology that helped Rienhoff arrive at his current hypothesis can be repeated to search for other potential causes for Bea's symptoms. Given Rienhoff's education, skills and resources, a diagnosis of Bea's specific disorder is within reach.

Rare disorders are not as rare as the label implies. They are defined simply as those affecting less than about 1% of the population. The US National Organization for Rare Disorders (NORD.org) estimates that there are around 6,000 rare disorders affecting 25 million Americans today. Thus, nearly one in 12 Americans suffer from a medical condition not addressed by blockbuster drugs (those with annual sales of over $1 billion) or mainstream medical research.

Rare diseases offer a clear example of the need to move away from the totalitarian-style "one drug fits all" model of drug development and to move towards individualized diagnosis and treatment. But personalized approaches to medicine are also revolutionizing the prevention and treatment of common disorders such as cancer, diabetes and heart disease. For example, Ashkenazi Jewish women with a history of cancer in their families are now routinely tested for any of three deleterious mutations in two genes called *BRCA1* and *BRCA2*.[2] Women who test positive have a greater than 85% probability of developing breast cancer during their lifetime.[3] Many choose to undergo preventive surgery. Genetic tests for these mutations have been available since the mid-90s[4] and are widely used.

About a quarter of women who are diagnosed with breast cancer harbor a genetic mutation that causes them to over-produce a protein called Her2. The over-production of Her2 leads to particularly fast-growing tumors. Today, women with breast cancer can take a test to check if they have the Her2 mutation. If they do, they can take a drug (Herceptin) which neutralizes the effects of the errant protein and directs the immune system to attack Her2-rich tumor cells.

---

[2] Due to patent restrictions, all the tests are performed by Myriad Genetics (http://www.myriad.com/products/bracanalysis.php). A full breakdown of the prevalence of breast and ovarian cancers for people of Ashkenazi and non-Ashkenazi background and various other family backgrounds is given at http://www.myriadtests.com/provider/brca-mutation-prevalence.htm. A full specification of the tests is given at http://www.myriadtests.com/provider/doc/tech_specs_brac.pdf.

[3] FH Fodor *et al.*, Frequency and carrier risk associated with common BRCA1 and BRCA2 mutations in Ashkenazi Jewish breast cancer patients, *Am. J. Hum. Genet.*, 1998, **63**: 45–51.

[4] MF Myers *et al.*, Genetic testing for susceptibility to breast and ovarian cancer: Evaluating the impact of a direct-to-consumer marketing campaign on physicians' knowledge and practices, *Genetics in Medicine*, 2006, **8**(6): 361–370.

Most breast cancer patients who do not have the Her2 mutation instead express another protein (ER) in their tumor cells. For these patients, the ER-specific drug Tamoxifen has been shown to be particularly effective. Using target-specific drugs such as Herceptin and Tamoxifen, combined with tests that identify a patient's particular molecular profile can dramatically improve the rate of recovery from breast cancer.[5]

Another way in which personalized medicine is already helping patients is in minimizing the adverse effects of drugs. In 1994, in the United States alone, 2.2 million people required hospitalization due to adverse drug reactions.[6] In the UK, adverse drug reactions accounted for 6.5% of all hospital admissions during 2001–2, at a projected annual cost of $847 million.[7] The US population is roughly five times that of the UK, so the corresponding annual cost in the USA would be more than $4 billion.

There are no drugs with absolutely no adverse effects. For most patients and drugs, the benefits far outweigh the adverse effects. But this is not true for everybody. For example, heart failure strikes about one in five adults, and kills about one in four of those affected within a year.[8] "β-blockers" can delay the need for a heart transplant significantly. But patients diagnosed with heart failure usually have to take a cocktail of half a dozen drugs every day, increasing the risk of drug interactions and adverse effects. It turns out that some 40% of African American males carry a genetic mutation that mimics the function of β-blockers. For these people, taking β-blockers has no benefit. It just adds to the risk of drug interactions and adverse effects.[9] Now that this particular mutation is known, African Americans diagnosed with heart failure can be genetically tested and only prescribed β-blockers if they do not have the mutation.

---

[5] K Altundag, FJ Esteva and B Arun, Monoclonal antibody-based targeted therapy in breast cancer, *Current Medicinal Chemistry — Anti-Cancer Agents*, 2005, **5**(2): 99–106. CS Sawyer, The cancer biomarker problem, *Nature*, 2008, **452**: 548–552.

[6] J Lazarou *et al.*, Incidence of adverse drug reactions in hospitalized patients, *JAMA*, 1998, **279**(15): 1200–1205.

[7] M Pirmohamed *et al.*, Adverse drug reactions as cause of admission to hospital: prospective analysis of 18,820 patients, *BMJ* **329**: 15–19.

[8] D Levy *et al.*, Long-term trends in the incidence of and survival with heart failure, *New England Journal of Medicine*, 2002, 347(18): 1397–1402. VL Roger, Trends in heart failure incidence and survival in a community-based population, *Journal of the American Medical Association*, 2004, 292(3): 344–350.

[9] SB Liggett, A GRK5 polymorphism that inhibits β-adrenergic receptor signaling is protective in heart failure, *Nature Medicine*, 2008, **14**(5): 510–517.

Why Now?

In a limited way, genetic testing and personalized medicine have been around for more than half a century. In the USA, since the 1960s newborn children have been tested for blood levels of the enzyme phenylalanine, whose absence can cause brain damage.[10] Children diagnosed with the Phenylketonuria (PKU) disorder can avoid brain damage by adopting a strict diet. Although PKU is a genetic disorder, the PKU test is not DNA-based. The first *genetic* test was developed for Huntington's disease.

Huntington's is a neurodegenerative disease caused by a mutation in a single gene (*Huntingtin*).[11] The mutation cannot be detected effectively in the blood. DNA-based tests for Huntington's were developed and made available to the public in the mid-80s. Since about the same time, the New York based Dor Yeshorim organization has been offering anonymous pre-marital genetic screening for (recessive) genetic diseases that affect children if both parents are carriers (such as Tay-Sachs). Currently, the organization tests for ten such disorders affecting Ashkenazi Jews.[12] Similar efforts are under way for other genetically well-defined populations such as the Old Order Mennonites and the Amish.

Figure 1.1 charts the growth (in the USA) of tests for genetic disorders over the past 15 years.[13] As of May 2009, tests for more than 1,700 conditions were available from over 600 laboratories in the USA.[13]

If genetic and personalized medicine have been around for so long, why are they suddenly receiving so much attention in the media? The short answer is biotechnology.

Fuelled by the success of the human genome project, our ability to sequence DNA has been increasing at an exponential rate. For example, figure 1.2 shows the cost per finished base pair from the earliest days of DNA sequencing to the present.[14] Similar gains have been made in throughput. The upshot is that the kind of genetic detective work that was the

---

[10] See http://tinyurl.com/Phenylketonuria for a description of Phenylketonuria (PKU) and references to literature.

[11] See http://www.hdsa.org/ and http://www.hdfoundation.org.

[12] http://en.wikipedia.org/wiki/Dor_Yeshorim; M Gessen, *Blood Matters: From Inherited Illness to Designer Babies, How the World and I Found Ourselves in the Future of the Gene*, Harcourt, 2008, Chapter 8.

[13] Source GeneTests: Medical Genetics Information Resource (online database). Copyright, University of Washington, Seattle. 1993–2009. Available at http://www.genetests.org, accessed May 2009.

[14] Figure kindly provided by Ray Kurzweil (personal communication). An earlier version of this figure appeared in Ray Kurzweil's *The Singularity is Near*, Penguin Books, 2005, p. 73. A copy of the original figure is available at http://singularity.com/charts/page73.html.

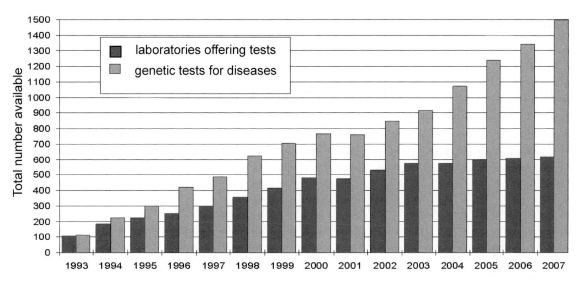

Figure 1.1: Growth of tests for genetic disorders in the US from 1993 to 2007.

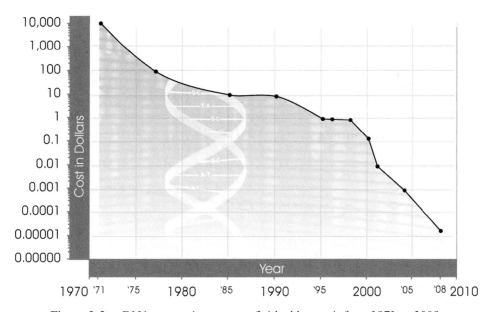

Figure 1.2: DNA sequencing cost per finished base pair from 1971 to 2008.

subject of entire PhDs in the late 1980s can now be performed by an undergraduate biology student as a summer project at less than 1/1000th of the cost.

Figure 1.3 shows the number of entries in the Human Gene Mutation Database[15] over the past 30 years. Note how the number of new entries per year is increasing every year. We are unmistakably at the start of a new era of genetic knowledge. Current projections suggest that we will know all mutations that occur in 1% or more of the human population within a few years.[16]

Developing a catalog of mutations and associated disorders is only part of the personalized medicine story. A second part of the story concerns the development of technologies that allow detailed molecular measurements of the health status of individual patients. Technological developments in the past decade have led to a veritable arsenal of tools that allow detailed quantitative measurements of cellular function and dysfunction throughout the body. Example tests already in regular use are colonoscopies (to detect colon cancer),

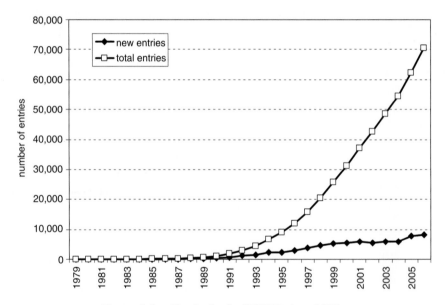

Figure 1.3: Entries in the HGMD since 1979.

---

[15] Data from http://www.hgmd.cf.ac.uk/; see PD Stenson *et al.*, The Human Gene Mutation Database: 2008 update, *Genome Medicine*, 2009, **1**: 13.

[16] See for example DE Reich, SB Gabriel and D Altshuler, Quality and completeness of SNP databases, *Nature Genetics*, 2003, **33**(4): 457–458.

testing for inactive TPMT alleles before cancer treatment with thiopurine drugs, bone density (DEXA) scans, and serial CA-125 blood testing (to detect ovarian cancer). Later in this book, we will review the impact of ultrasound, MRI, PET and CAT scans, and a large variety of molecular measurements using bodily fluids.

The third part of the personalized medicine story is about our ability to interpret the data from genomic and health-monitoring assays. The genetic success stories discussed earlier in this chapter either relate to single-gene disorders (e.g. Huntington's) or disorders for which a combination of a single-gene mutation and personal/family history is highly predictive (e.g. *BRCA1* for breast and ovarian cancer). Unfortunately, the great majority of genetic effects on diseases involve interactions among multiple mutations and a large variety of environmental and life-history factors. As a result, it has been difficult to predict the outcome of most mutations.

To illustrate the problem, consider figure 1.4. Currently, some 268 mutations are thought to contribute to Alzheimer's disease.[17] A small fraction of these genes are shown in the figure (nodes labeled with gene names). The underlined nodes represent the disorders caused by mutations in the genes they are connected to (the arrows link gene mutations to diseases).

For the purposes of our discussion here, the names of the particular disorders and genes shown do not matter. Note instead how most genes contribute to multiple disorders. Not all the disorders caused by Alzheimer-related mutations are neurological. In the figure, neurological and cardiological disorders are indicated. 40% of the genes shown contribute to both disorders.

The principles underlying the above observations have been known for a long time. A single gene will take part in multiple biological processes throughout the body. So it is not surprising that a mutation in a single gene may affect multiple biological functions. At the same time, any single biological process is the outcome of highly regulated interactions among a large number of genes, so a mutation in any single gene will have a complex, context-dependent effect on the outcome.

The upshot is that until recently it has not been possible to predict the health outcomes of most genetic mutations and environmental factors. The emergence and maturation of integrative, network-oriented, and whole-genome approaches over the past decade is changing that. As the rest of this book will show, we now have the technologies and methodologies

---

[17] Data retrieved June 5th 2008 from the Online Mendelian Inheritance In Man (OMIM) database: http://www.ncbi.nlm.nih.gov/omim. The network shown is a small portion of the OMIM Morbid Map visualized in Cytoscape (http://www.cytoscape.org/).

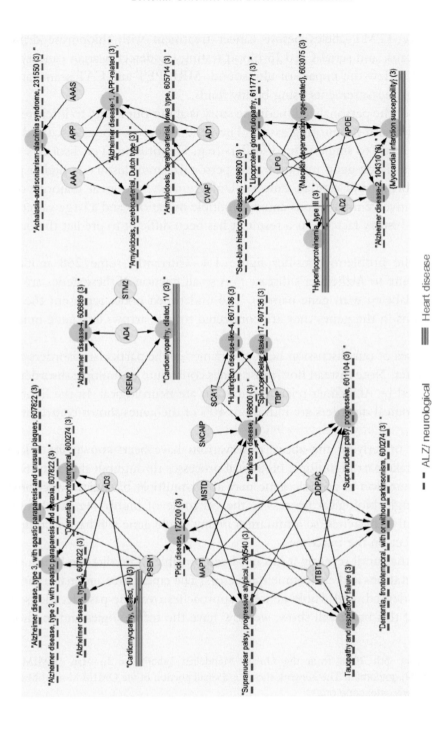

Figure 1.4: Some genes associated with Alzheimer's disease in OMIM.

necessary to understand how interactions between a person's genome and environmental exposures may lead to cellular and organ dysfunction, to develop customized treatments for identified dysfunctions, and to monitor and control the progress of treatments closely.

Biochemical Individuality: The Need for Personalized Medicine

The above discussion provides some specific examples of personalized medicine in action. But one might argue that medicine has always been inherently personal. So what is different now? Isn't it simply that better diagnostic/health-status tests are delineating diseases more specifically and stratifying diseases into sub-types? For example, autism is no longer viewed as a single disease, but as a spectrum of related developmental disorders. This is certainly the general direction in which medicine is moving. But, as discussed below and in subsequent chapters, there are also several reasons to think that *each patient* presents a unique condition and will ultimately be best served by a fully individualized treatment plan.

The concept of biochemical individuality goes back more than a century. When Archibald Garrod — one of the founding fathers of clinical genetics — published his breakthrough paper in 1902, he gave it the subtitle "A study in chemical individuality".[18] In his 1956 book *Biochemical Individuality*,[19] the biochemist RJ Williams catalogued inter-individual differences in organ size and shape, physiological measurements, numbers of various immune cell types, and a range of metabolic and other biochemical parameters. For example, he noted that the relative abundances of some immune cells in the bone marrow vary by more than 20-fold between apparently healthy individuals.[20]

In addition to genetic differences among individuals, there appear to be three primary causes of biochemical individuality. Firstly, the human digestive system relies heavily on a large variety of commensal (good, symbiotic) bacteria whose relative abundances differ greatly from person to person (even between identical twins[21]). Approximately 200 different phyla of commensal bacteria are currently thought to occupy the colon.[22] Each person tends

---

[18] AE Garrod, The incidence of Alkaptonuria: A study in chemical individuality, *The Lancet*, 1902, **ii**: 1616–1620. Available from the Electronic Scholarly Publishing Project: http://www.esp.org/foundations/genetics/classical/ag-02.pdf.

[19] RJ Williams, *Biochemical Individuality — The Basis for the Genetotrophic Concept*, originally published in 1956, re-published by Keats Publishing in 1998.

[20] Ibid., pp. 35–37.

[21] PJ Turnbaugh *et al.*, A core gut microbiome in obese and lean twins, *Nature*, 2009, **457**: 480–484.

[22] L Dethlefsen *et al.*, An ecological and evolutionary perspective on human–microbe mutualism and disease, *Nature*, 2007, **449**: 811–818; A Mullard, The inside story, *Nature*, 2008, **453**: 578–580.

to have different species and numbers of bacteria in each phylum.[23] The various bacterial groups interact with and cross-regulate each other as well as the gut epithelium.[24] Differences in microbiome composition appear to have important health implications. For example, obese and lean individuals have distinctly different microbiome compositions, and their microbiomes express large numbers of genes differentially.[25]

Figure 1.5 shows the relative abundance (arbitrary scale) of two common families of gut bacteria in two infants.[26] The dashed lines represent the first microbial family, the solid lines the second. The horizontal axis indicates the number of days since birth. Note both the dramatic difference between the two infants (compare same-type curves in the two panels), and also the large degree of variability over time.

The exact balance of bacterial species in each individual also depends on cross-regulatory interactions with the host immune system.[27] Overall, the digestive and metabolic systems are

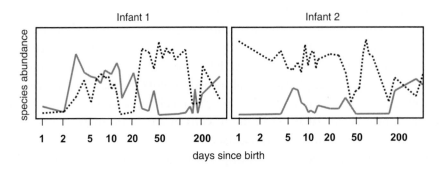

Figure 1.5:   Abundance of gut bacteria in two infants.

---

[23] A Mandavilli, Straight from the gut, *Nature*, 2008, **453**: 581–582.

[24] MA Mahowald *et al.*, Characterizing a model human gut microbiota composed of members of its two dominant bacterial phyla, *Proceedings of the National Academy of Sciences of USA*, 2009, **106**(14): 5859–5864.

[25] See Ref. 21 and also PJ Turnbaugh *et al.*, An obesity-associated gut microbiome with increased capacity for energy harvest, *Nature*, 2006, **444**: 1027–1031; RE Ley *et al.*, Human gut microbes linked to obesity, *Nature*, 2006, **444**: 1022–1023.

[26] Figure adapted from C Palmer *et al.*, Development of the human infant intestinal microbiota, *PLoS Biology*, 2007, **5**(7): e177.

[27] S Mazmanian, A microbial symbiosis factor prevents intestinal inflammatory disease, *Nature*, 2008, **453**: 620–625; MGH Besselink *et al.*, Probiotic prophylaxis in predicted severe acute pancreatitis: a randomised, double-blind, placebo-controlled trial, *The Lancet*, 2008, **371**(9613): 651–659.

further individualized by the fact that they interact extensively with the immune and auto-nomic nervous systems.[28]

The second major cause of inter-individual differences in disease susceptibility and drug-response is the immune system itself. Developmental history, as well as the frequency and type of exposures to diseases during childhood, greatly affect the repertoire and pre-dispositions of the immune system. For example, susceptibility to allergies can be reduced if an infant is reared on mother's milk,[29] and different types of childhood infections can increase or decrease the probability of allergies in later life.[30] As noted earlier,[20] the rela-tive abundance of some immune cells (e.g. eosinophils) in healthy individuals can vary enormously.

The third major cause of inter-individual differences in disease susceptibility and response to treatment is environmental exposure to toxic materials (reviewed in Chapter 4), as exem-plified by the effects of cigarette smoke, alcohol, and hallucinogenic drugs. While exposure to cigarettes, alcohol and drugs may be voluntary, other exposures may be unintentional. For example, the role of endocrine disruptors — chemicals that mimic the action of hormones — in birth defects, infertility and cancer[31] has been well publicized.[32]

The Demand for Individualized Medicine

The preceding discussions argued that personalized medicine would be beneficial and that the technologies that would permit widespread personalized medicine are now becoming sufficiently cheap and effective. But is there strong demand for personalized medicine, or is it just marketing hype?

There are several reasons to think that personalized medicine will be highly sought after. A key factor is that the average age in industrialized countries is increasing. Figure 1.6 shows

---

[28] MD Gershon, *The Second Brain*, 1998, HarperCollins. Reprinted as a Quill paperback in 2003.

[29] V Verhasselt *et al.*, Breast milk-mediated transfer of an antigen induces tolerance and protection from allergic asthma, *Nature Medicine*, 2008, 22: 170–175.

[30] JE Gern and WW Busse, Relationship of viral infections to wheezing illnesses and asthma, *Nature Reviews Immunology*, 2002, 2: 132–137; M Jackson, *Allergy: The History of a Modern Malady*, Reaktion Books, 2006, Chapter 5.

[31] F Ohtake *et al.*, Dioxin receptor is a ligand-dependent E3 ubiquitin ligase, *Nature*, 2007, **446**: 562–566; J Wade-Harper, A degrading solution to pollution, *Nature*, 2007, **446**: 499–500.

[32] For a balanced in-depth review, see the 2002 World Health Organization report: Global assessment of the state-of-the-science of endocrine disruptors. Available online at http://www.who.int/ipcs/publications/new_issues/endocrine_disruptors/en.

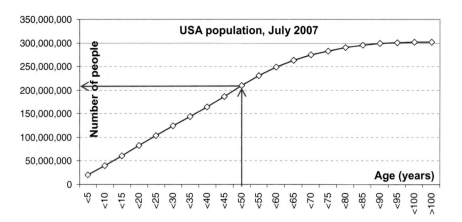

Figure 1.6:   Population of USA in 2007 by age.

the population of the USA by age.[33] The total population, in July 2007, was estimated to be just over 300 million. As indicated by the arrows in the figure, about one third of this population is over 50. A similar picture is seen in Western Europe and Japan. For example, in the UK about 16% of the population is more than 65 years old.[34]

Older populations tend to require more medical care. Importantly, older people have a more diverse health history and therefore more varied healthcare needs. They are also more prone to drug adverse effects because they are more likely to have multiple ailments (and concomitant treatments). For these reasons, individualized diagnosis (including genomics) and preventive medicine will be in increasing demand by a large proportion of people in industrialized countries.

Among younger people too, individualized diagnosis and treatment is becoming increasingly attractive as we discover more about the effect of genetic background on disease susceptibility and response to treatment. This is true not only for people from fairly homogeneous genetic backgrounds (e.g. Ashkenazi Jews), but also for those with mixed genetic heritage because greater mixing of previously isolated populations can potentially dissociate protective mutations from their targets (as has been hypothesized for type 1 diabetes[35]).

---

[33] Data from the US Census Bureau, at http://www.census.gov/popest/national/asrh/NC-EST2007-sa.html.

[34] Data from the UK Office for National Statistics, http://www.statistics.gov.uk/cci/nugget.asp?ID=949.

[35] ZL Awdeh *et al.*, A genetic explanation for the rising incidence of type 1 diabetes, a polygenic disease, *Journal of Autoimmunity*, 2006, **27**: 174–181.

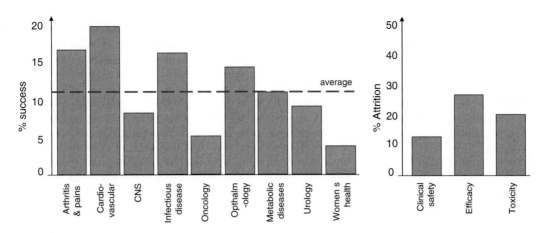

Figure 1.7: Success and attrition rates during clinical trials (1991–2000).

On the supply side, personalized medicine is also highly attractive to the pharmaceutical industry. Two important reasons for this are highlighted in figure 1.7.[36]

The left panel shows the success rate during clinical trials for various types of disorder. The data are from ten large pharmaceutical companies and cover the period 1991–2000. As indicated by the dashed line, the average success rate is only about 11%. For some disease areas, such as women's health, Central Nervous System (CNS) and cancer (oncology), the success rates are even lower. The right hand panel shows three of the four top causes of attrition in the year 2000: safety, efficacy and toxicity. Together, these factors account for about 60% of failures. As discussed earlier, efficacy, safety and adverse-effect issues can be improved through better matching of drugs to patients' specific genetic make-up and medical history.

Beyond success in clinical trials, personalized medicine can also help drug manufacturers avoid the boom and bust cycle of blockbuster drugs. When large-scale use of a blockbuster drug turns up adverse effects, the predicted loss of income can result in a large and rapid drop in the parent company's stock price. For example, when in 2007 GlaxoSmithKline's blockbuster drug for type 2 diabetes (Avandia) was reported to increase the risk of heart attacks,[37]

---

[36] Reprinted with permission from Macmillan Publishers Ltd: I Kola and J Landis, Can the pharmaceutical industry reduce attrition rates? *Nature Reviews Drug Discovery*, 2004, **3**: 711–715. Copyright (2004).

[37] SE Nissen and KN Wolski, Effect of Rosiglitazone on the risk of myocardial infarction and death from the cardiovascular causes, *New England Journal of Medicine*, 2007, **356**(24): 2457–2471.

US sales of Avandia dropped by 48% in a single quarter.[38] GlaxoSmithKline is one of the largest pharmaceutical companies in the world, yet its share price fell by 13% during this period.[38] Such fluctuations in share prices make it difficult for companies to plan long-term. Personalized drugs would allow companies to provide a wider variety of drugs, each targeted to a smaller, more specific category of patients. In this way, adverse effects on both patients and companies can be minimized. Moreover, drugs with well-defined and highly specific modes of action can be combined in patient-specific ways to provide better efficacy (we will return to this idea in the final chapter). Pharma can benefit from having a broad portfolio of widely used drugs, while patients benefit from customized drug combinations targeted to their exact needs.

Thus the development of personalized medicine is being driven by a remarkable confluence of supply and demand. Personalized diagnostics and treatment will prevail because — remarkably — patients, bio-medical researchers, physicians, medical-device manufacturers, and pharmaceutical companies will all benefit from it.

Personalized Medicine: The Vision

Imagine it is 2015. The genomes of thousands of individuals have been sequenced and compared. From these studies a catalog of (nearly) all viable variations[39] in human DNA sequences has been constructed. Genetically similar sub-populations have been identified and statistically compared. Specific models of genetic and environmental contributions to common and rare diseases have been developed.

At the same time, the development of sensitive, low-cost and non-invasive new tests will allow physicians to characterize a patient's status in quantitative biochemical detail. At regular check-ups, the physician will be able to screen the patient for early signs of any diseases to which the patient may be genetically predisposed. She may use a hand-held device to prick the patient's finger and measure the abundance of any of thousands of organ-specific biomarker proteins that diffuse into the blood stream from all organs. To this she may add observations from the patient's urine, saliva and stool, various body-part scans, and specific diagnostic assays. The data are sent via wireless communication to a server, which in turn will analyze the information and email a report back to the patient and the physician.

Next, the patient's symptoms, genomic data, environmental exposure history, and blood-borne biomarker data are used to arrive at specific hypotheses regarding the

---

[38] B Vastag, Reviewer leaked Avandia study to drug firm, *Nature*, 2008, **451**: 509.

[39] i.e. mutations that do not result in abortive embryonic development.

molecular causes of the symptoms presented. Each hypothesis is tested experimentally with more stringent diagnostics, and targeted molecular probes that validate or falsify it. Iterative refinement of hypotheses leads to a specific diagnosis of dysregulated genes and pathways in the patient.

Because patients can be monitored on a regular basis for given predispositions, diseases can be detected early on and treated as dynamic, changing processes. Routine, detailed, quantitative and personal diagnostics will serve three critical purposes. First, a disease that is detected early enough can often be cured completely (e.g. for many cancers). If a cure is not possible, early detection will allow better management and containment of the disorder (e.g. drugs that delay the onset of Alzheimer's disease). Second, better disease stratification to identify the type and stage of progression of the disease will allow better treatment decisions. For example, men with prostate cancer are able to choose between various forms of radiation therapy, surgery, or "watchful waiting". Depending on their age, the stage of the cancer, and other factors such as family considerations, they may choose a balance between treatment with adverse effects (e.g. incontinence and impotence) and the risks of progression.[40]

The third benefit of routine, molecularly detailed, and quantitative personal health monitoring is in the insights that will arise from analysis of de-identified medical records. As we will see later in this book, patient records are increasingly held electronically. A variety of patient-advocacy groups, health-providers, and health-informatics suppliers are already exploring aggregated health records to identify trends, divide diseases into sub-types, and track the long-term effects of changes in patient-care procedures. Assuming patients can be well-protected from privacy breaches, the advent of personalized genomics and medicine will provide molecularly detailed longitudinal data from thousands to millions of patients with similar symptoms. In effect, each patient will be able to draw on the experiences all of the other individuals with similar symptoms. Moreover, researchers analyzing past records may discover novel correlations leading to new, testable hypotheses and treatments. Some patients with particularly debilitating conditions may elect to become their own ultimate "model organism", volunteering to test new treatments developed for their particular genomic background and symptoms. The legal, moral, and societal challenges of these trends will be discussed in Chapters 10 and 11.

To return to our vision of a patient in the year 2015, what can a patient expect after a detailed diagnosis? At a minimum, the physician can use various databases to explore treatment

---

[40] See for example http://www.cancer.gov/cancertopics/prostate-cancer-treatment-choices.

options for the particular combination of disorders identified. Potential adverse interactions with the patient's genomic make-up, her environment, and her other drugs (for other disorders) can be characterized and weighed against the projected benefits/adverse effects of alternative treatments. By measuring the rate at which a patient's condition is deteriorating or improving, the physician can fine-tune the treatment plan.

The early-diagnosis nature of sensitive measurements and quantitative molecular models of clinical outcomes necessarily makes personalized medicine participatory. A physician can describe the alternative interpretations of a patient's condition, the different treatment options, the predicted rate of recovery in each case, etc. A counselor may be employed to help the inexpert patient understand all this information and ensure that the patient makes well-informed choices. But in the end, it is the patient who has to decide. The more detailed and quantitative medicine becomes, the more each patient becomes unique and so the greater the need to decide on the most appropriate course of action personally. We will explore the changing role of the patient further in Chapter 10.

The Aims of this Book

This book explains how a slew of recent biological, technological and methodological developments are making it possible to provide personalized diagnosis and treatment for every child and adult suffering from any disorder. Personalized medicine draws on such a broad spectrum of scientific disciplines that very few people currently understand every aspect of it. Yet, by virtue of treating each person's condition as unique, personal genomics and personalized medicine *require* that both physicians and patients understand the nature of the data, its health implications, and its limitations. This book aims to equip readers with the necessary knowledge to research, understand, and exploit personal genomics and personalized medicine.

Who Should Read this Book

Personal genomics and personalized medicine are already practicable realities. Within the coming decade, their use will become widespread. There is therefore a pressing need for current biomedical students and researchers, physicians, genetic counselors, and advocacy groups to learn about personalized genomics and medicine. Yet, because of the multidisciplinary nature of personalized medicine, few professionals have all the necessary expertise. This book can serve as a primer in personal genomics and personalized medicine for practicing biomedical researchers, health professionals and students.

Personal genomics and personalized medicine are already available, demanding informed decision-making by patients/clients. Thus, in a very real sense, this book is also for anyone who might be diagnosed with a major medical condition in the near future.

Organization of the Book

The chapters of this book are arranged so as to present a progression of concepts. The earlier chapters review the biological and technical topics. The later chapters assume the reader is familiar with these topics, and focus on practical issues relating to the implementation of personal genomics and personalized medicine.

Chapter 2 provides a brief overview of molecular biology and related topics essential to this book. The reader is assumed to already have an education in these subjects, and the review presented here is intended only to provide a refresher and quick reference for the discussions in the rest of this book.

Chapter 3 reviews the sources, types and prevalence of DNA sequence differences among individuals. We see that genetic differences between individuals are widespread, and cause differences in cellular processes and disease susceptibility.

Chapter 4 reviews how environmental factors interact with an individual's genetic make-up throughout life, leading to both short-term and lasting effects on cellular and organ function.

Chapter 5 discusses how the genetic and environmental influences presented in Chapters 3 and 4 lead to variations in the biochemical function and behavior of cellular pathways. We see that different types of variation affecting the same molecule can have very different effects depending on the context, and note the need to characterize inter-individual variability in terms of the effects of DNA sequence variability and environmental factors on specific cellular pathways and their role in organ physiology.

Chapters 6 and 7 review the emerging technologies for DNA sequence analysis and for molecular measurement of cellular and organ function. We see that a wide variety of emerging technologies are making it possible to sequence the entire genome of an individual for under $1,000, while other technologies are providing increasingly cheap, comprehensive, quantitative, and non-invasive tools for inspecting the molecular health of individuals on an ongoing basis.

Chapter 8 presents current approaches to the interpretation of genetic and health-monitoring data. While many of the tools being developed for these purposes are currently in their infancy, we note that a clear path forward is already well delineated and discuss the way in which these tools will change the nature of medicine in the coming decade.

Chapter 9 explores how personal genomics and biomarker data can be used in clinical practice. In particular, we will look at the use of computational resources that facilitate access to and interpretation of genomic and biomarker data. We also note that large-scale adoption of electronic health records and other IT infrastructure will accelerate "learning from practice" approaches.

Chapter 10 looks at the many social, ethical and legal issues surrounding personal genomics and personalized medicine. Finally, Chapter 11 provides a summary of the steps we will need to take *now* to ensure effective exploitation of personal genomics and personalized medicine in the near future.

How to Read this Book

The chapters in the book are ordered to provide a progression of ideas. For teaching purposes, I hope that the order of the chapters will provide a natural sequence of lectures.

For those who are not reading this book as part of a course, the multi-disciplinary nature of the subject matter makes it likely that any one reader will already be well-versed in *some* of the topics presented. I would like to encourage such readers to sample chapters and read those that interest them most first. To facilitate non-linear paths through the book, I have tried to provide ample cross-references to other chapters.

I have also tried to organize the book such that it can be read at two different levels. In each chapter, extensive footnotes provide references and points of departure for further reading. Readers wishing to delve more deeply into a topic are encouraged to check the footnotes and references as they read the main text.[41] Readers looking for an introductory overview can read the main text and ignore the footnotes.

Acknowledgments

This book came about largely because of my many discussions about personal genomics and personalized medicine with Lee Hood (Institute for Systems Biology, Seattle, USA). My discussions with Lee started when I joined the faculty of the ISB in 2002 and they became

---

[41] Some very long or machine-generated URLs have been abbreviated using the (free) service provided by http://tinyURL.com. Inputting such URLs into a web browser will redirect you to the full address.

ever more regular and intensive after I left the ISB at the end of 2005. Invariably, Lee would ask me what I would do if I were presented with thousands of personal genomes that very day, and I would come away from our meeting with the realization that there was much I needed to learn.

This book is in essence a report of what I learnt as I tried again and again to answer Lee's question more fully. I am deeply grateful to Lee for his encouragement and advice. In particular, at the annual Hood lab retreat at Lee and Valerie's ranch in Montana in July 2008, Lee kindly spent an entire afternoon commenting on my initial chapter outlines for this book. Many thanks, Lee.

For my continuing training in molecular biology and immunology, I would like to thank my long-time collaborators Eric Davidson and Ellen Rothenberg (Division of Biology, California Institute of technology). Also, in 2006, I spent a very enjoyable and instructive year in the laboratories of Alan Aderem and Adrian Ozinsky (Institute for Systems Biology, Seattle) learning how to perform a variety of single-cell assays, especially using fluorescent microscopy and microfluidics. I am particularly grateful to Adrian Ozinsky, April Clark, Heather Kostner, and Alan Diercks, for spending many patient hours guiding me. Thank you.

Special thanks to the following individuals for giving me their time, attention and infinitely valuable feedback on various chapters and iterations of this book:

Kas-Ray Badiozamani, MD
Radiation Oncology,
Virginia Mason Medical Center
Seattle, WA

Eric Davidson, PhD
Division of Biology
California Institute of Technology
Pasadena, CA

John Halamka, MD, PhD
CareGroup Health System &
Harvard Medical School
Boston, MA

Hugh Rienhoff, MD
MyDaughtersDNA.org

Anthony Blau, MD
Division of Hematology
University of Washington
Seattle, WA

Gustavo Glusman, PhD
Institute for Systems Biology
Seattle, WA

Jim Karkanias, PhD
Microsoft Health Solutions
Redmond, WA

Jared Roach, MD, PhD
Institute for Systems Biology
Seattle, WA

Lee Rowen, PhD
Institute for Systems Biology
Seattle, WA

Kelly Smith, MD, PhD
Department of Pathology
University of Washington
Seattle, WA

The production of a book is very much a team effort. I thank my production team collectively for their amazing efficiency. In particular, my publisher Laurent Chaminade and editors Lizzie Bennett and Xiao Ling have been paragons of friendly efficiency and a pleasure to work with.

Last and most important to me personally, I could not have written this book without the constant warmth and support of my partner Cecilia Bitz. Thank you, Cecie.

CHAPTER 2

From DNA Sequence to Physiology

The aim of this chapter is to introduce and define the biological terms and concepts that underlie subsequent discussions in the book. We assume the reader is already familiar with the basic concepts of molecular and cell biology[1] and focus on presenting a compact overview of those aspects of genomic and cellular organization that impinge on our later discussions of personal genomics and personalized medicine.

Genetic Information Flow

The classical paradigm of information processing in cells is that cellular (and hence organ and organism) form and function arise from the decoding of the information stored in inherited DNA. In this model, information in the DNA is encoded digitally in the form of the four nitrogenous bases adenine, cytosine, guanine and thymine (abbreviated to A, C, G, and T respectively). These bases are attached to a helical backbone of repeating sugar-phosphate molecules. The DNA double helix is formed by hydrogen bonding between "complementary" bases (A with T and C with G), giving rise to the complementary, anti-parallel, "Watson–Crick" DNA strands. Consecutive triplets of bases along a DNA strand encode the 20 amino acids that make up proteins. According to the Central Dogma, information encoded as base pairs in the DNA double helix is first transcribed into RNA molecules and then translated into proteins, which are considered the active agents of cellular form and function.

Over the past half century, this classical model of information processing has been revised extensively. Cellular information flow is no longer seen as a single linear chain. In this chapter, we will review some of the many ways in which information flow in cells is now considered to involve many feedbacks and lateral interactions among multiple information processes.

---

[1] See for example JD Watson *et al.*, *Molecular Biology of the Gene*, 6th Edition, 2008, Cold Spring Harbor Press; AJ Courey, *Mechanisms in Transcriptional Regulation*, 2008, Blackwell; B Alberts *et al.*, *Molecular Biology of the Cell*, 4th Edition, 2002, Garland Press; SF Gilbert, *Developmental Biology*, 8th Edition, 2006, Sinauer Associates.

In particular, we will discuss the roles of the cellular and organismic environment on cellular development and function. As we will see, cellular information processing is highly nonlinear and heterogeneous across cell types and over time.

All the cells of the adult body are descendants of the fertilized egg (zygote). Paternal DNA is injected into the maternal egg by a single sperm cell. Apart from this, almost all the contents of the egg are inherited from the mother. In particular, zygotic mitochondria — which possess their own 16.5 Kb DNA and are essential components of eukaryotic cellular metabolism — are maternally inherited.[2]

The ~6 billion base pairs of human DNA are divided into 23 pairs of chromosomes. Apart from the sex chromosomes (X and Y), chromosomes 1 to 22 from each parent are homologous. Germ cells undergo meiosis to produce haploid sperm and egg cells. Crossover (recombination) during meiosis mixes paternal and maternal chromosomes so that each egg and sperm cell is genetically unique.

After crossover, the chromosomes are repeatedly copied as cells divide to give rise to the body. In addition to crossover a variety of DNA replication "errors" can generate variation in the offspring DNA. Moreover, over the lifetime of a cell, many processes can chemically modify DNA nucleotides and chromatin in heritable ways. An overview of the sources and types of DNA variation is presented in the next chapter.

Developmental processes during the first few cleavages are driven by maternally inherited factors. In humans, zygotic transcription starts at around the third cleavage and gradually takes over the regulation of cellular processes.[3] Because mammalian embryos grow in the highly specialized environment of the uterus, they can receive many developmental signals (as well as nutrients) from the mother. Thus, mammalian embryonic development is not a self-contained process, but one that is regulated maternally. The upshot is that environmental factors can influence the interpretation and usage of inherited genetic information from the very earliest stages of embryonic development. For example, the children of mothers who gain too much weight during pregnancy are more likely to be obese at three years of age[4] (after corrections for all suspected confounding factors).

The early cell divisions generate two distinct groups of cells: an outer layer of cells that eventually become part of the placenta, and an inner cell mass of embryonic stem cells (ESCs)

---

[2] We will return to this issue when we discuss inheritance of mitochondrial genetic diseases in later chapters.

[3] P Braude, V Bolton and S Moore, Human gene expression first occurs between the four- and eight-cell stages of preimplantation development, *Nature*, 1988, **332**(6163): 459–461.

[4] E Oken *et al.*, Gestational weight gain and child adiposity at age 3 years, *American Journal of Obstetrics and Gynecology*, 2007, **196**(4): 322.e1–e8.

that will eventually give rise to all the cells of the body. Around one fifth of the 64 cells of the sixth cleavage embryo are inner cell mass ESCs. Monozygotic (identical) twins are usually the result of the division of ESCs into two separate embryos in this early phase (within the first nine days of development).

Germline cells differentiate from somatic cells between embryonic weeks 2 and 3, just before gastrulation. Shortly afterwards, the primordial germs cells migrate to the outside edge of the embryo (from where they will eventually migrate to the gonads). Sperm cells in adult males are continually derived from a population of germline stem cells in a process that takes over two months. In contrast, all of the eggs ovulated during the lifetime of a woman are "pre-manufactured" before birth.

The separation of germline cells from somatic cells is important for our discussions in this book because it means that germline cells are buffered from many of the subsequent cellular history and environmental influences on genomic material in somatic cells.

DNA Organization

The linear sequence of DNA is inherently directional. By convention, the carbon atoms in the sugar ring that is part of each nucleotide are numbered sequentially. Successive nucleotides in DNA form covalent bonds between the phosphate group attached to carbon 5 and the hydroxyl group attached to carbon 3. These nucleotide linkage positions are referred to as the 5′ (five prime) and 3′ ends. RNA transcription and DNA synthesis are performed by starting at the 5′ end of the nascent copy and adding new nucleotides to the 3′ end. Panel (A) in figure 2.1 schematically illustrates the arrangement of bases on the sugar backbone, and the phosphate linkage between successive sugar molecules. The DNA double helix is composed of two complementary strands of this structure, as schematically illustrated in panel (B).

In vivo, DNA is usually packaged into chromatin, an approximately 50:50 DNA-protein complex. The building block of chromatin is the nucleosome (panel C). Nucleosomes comprise about 150 nucleotides of DNA wrapped around a complex of eight histone molecules. Histones are small proteins with long "tails" that extend outside the octamer complex and can be chemically modified in many different and specific ways to regulate chromatin state and gene transcription.

Successive nucleosomes are connected together via ten to 50 nucleotides of linker DNA and an associated linker histone. In a process thought to be mediated by histone-histone interactions, nucleosomes pack together in a compact manner to form euchromatin fibers (panel D). Transcriptionally inactive regions of DNA, and the entire chromosome during cell

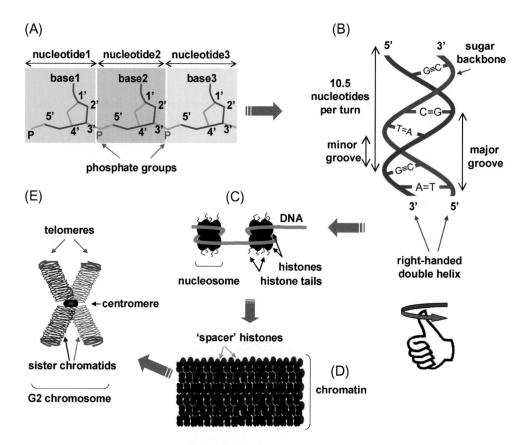

Figure 2.1:   DNA structure and organization.

division, are further condensed into a more tightly packed form of chromatin known as heterochromatin (panel E).

Less than 2% of the human genomic DNA codes for exons (protein-coding sequences).[5] The vast majority of our DNA sequence codes for features that facilitate functions such as:

- **DNA evolution.** For example, many varieties of short and long repeat sequences, including transposable elements, can be highly mutagenic. Sources and types of variation in DNA will be discussed in the next chapter.

---

[5] MK Sakharkar *et al.*, An analysis on gene architecture in human and mouse genomes, *In Silico* Biology, 2005, **5**(4): 347–365.

- **DNA replication, maintenance and assembly.** Example structures include chromosomal centromeres and telomeres, origins of replication, and histone-binding sequences.

  DNA replication is a complex, multi-step process involving many dedicated enzymes. In some ways it is similar to RNA transcription, which will be reviewed in the next section. A key difference is that mammalian DNA is copied simultaneously at many origins of replication. Some DNA replication errors will be reviewed in Chapter 3. Here, we note that errors in DNA replication, assembly and maintenance underlie many diseases. For example, origins of replication are present every few hundred bases in the human genome.[6] Consequently, aberrant formation of protein complexes at origins of replication can interfere with the formation of gene regulatory complexes. This form of interference underlies the inactivation of the p15/p16/ARF tumor suppressor genes.[7]

  Human centromeres are protein-DNA complexes containing several megabases of a 171bp repetitive DNA sequence containing a binding site for a histone-variant protein.[8] The protein-DNA complex forms a rigid structure which binds sister chromatids together during cell division, and marks the chromosomal location of the centromere in daughter cells.[9] Improperly located or malfunctioning centromeres can result in chromosomes that do not align or separate properly. For example, centromere malfunction can lead to aneuploidy, in which a daughter cell ends up with an incorrect number of chromosomes (as in tumors and Down syndrome, which involves a trisomy of chromosome 21).

  Telomeres are protein-DNA complexes formed at 6bp repetitive DNA sequences (TTAGGG)$_n$ at the ends of each chromosome. Human telomeres are typically thousands of base-pairs long. They protect the ends of chromosomes from degradation and stop chromosomes from fusing end-to-end. In most somatic cells, telomeres are shortened at each cell division, which can limit the number of times a somatic cell may divide. In germ cells (and some somatic cells), telomere lengths are actively restored after each cell division.[10]

---

[6] YJ Machida, JL Hamlin, and A Dutta, Right place, right time, and only once: replication initiation in metazoans, *Cell*, 2005, **123**: 13–24.

[7] S Gonzalez *et al.*, Oncogenic activity of Cdc6 through repression of the INK4/ARF locus, *Nature*, 2006, **440**: 702–706.

[8] PE Warburton, JS Waye, and HF Willard, Nonrandom localization of recombination events in human alpha satellite repeat unit variants: implications for higher-order structural characteristics within centromeric heterochromatin, *Mol. Cell. Biol.*, 1993, **10**: 6520–6529.

[9] BE Black *et al.*, Structural determinants for generating centromeric chromatin, *Nature*, 2004, **430**: 578–582.

[10] RJ Hodes, Telomere length, aging, and somatic cell turnover, *Journal of Experimental Medicine*, 1999, **190**(2): 153–156.

Cancer cells hijack this capacity to achieve unlimited replicative potential.[11] Accelerated telomere shortening caused by chronic stress[12] and inherited genetic disorders[13] have both been found to adversely affect the immune system through T-cell senescence.

- **Control of gene expression.** What, when and how much a gene is expressed is regulated in a multitude of ways. Apart from transcription factor genes, example regulatory sequences include promoters (sequences proximal to the transcription start site which bind factors that regulate the basal transcription apparatus), enhancers (potentially distant sequences to which regulatory transcription factors bind), non-coding RNAs (which often regulate levels of gene expression), and introns (intervening sequences between exons, which can regulate alternative splicing). These features are discussed in the next two sections, where we review transcription and its regulation.

An Overview of Transcription and Translation

The transcription of DNA into RNA is a complex process involving many dozens of gene products. The process may be broken into a series of steps, as follows[14]:

- Multiple regulatory factors interact to de-condense and open up heterochromatin, bend DNA, and recruit and activate RNA polymerase and associated proteins. This process is described in more detail in the next section.
- RNA polymerase and the general transcription factors are recruited to the promoter region. Together, these factors form the basal transcription apparatus (BTA). The BTA separates about ten bases of the DNA double-strand at the transcription start site and begins transcription.
- A pre-mRNA molecule is synthesized one nucleotide at a time. At each step, an RNA nucleotide capable of pairing with the current DNA nucleotide is added to the 3′ end of the nascent RNA transcript, then the transcription complex is moved one nucleotide along the DNA and the next nucleotide is copied. Transcription always proceeds in this

---

[11] Y-S Cong, WE Wright and JW Shay, Human telomerase and its regulation, *Microbiol. Mol. Biol. Rev.*, 2002, **66**(3): 407–425.

[12] AK Damjanovic *et al.*, Accelerated telomere erosion is associated with a declining immune function of caregivers of Alzheimer's disease patients, *Journal of Immunology*, 2007, **179**: 4249–4254.

[13] M Knudson *et al.*, Association of immune abnormalities with telomere shortening in autosomal-dominant dyskeratosis congenita, *Blood*, 2005, **105**(2): 682–688.

[14] The illustrative description here focuses on sequences coding for mRNA, rather than ribosomal or transfer RNAs which use different RNA polymerases and associated factors.

direction and only one strand of the double helix is used for transcription (specified by the promoter sequence). If an incorrect nucleotide is incorporated, the transcription complex moves backwards. The 3′ end of the nascent RNA dissociates from the DNA and is cleaved. Then transcription resumes.

- After the first 20 to 30 nucleotides have been transcribed, the dangling 5′ end of the transcript is capped with a modified nucleotide. The cap protects the RNA against degradation. It also aids subsequent mRNA transport out of the nucleus, and supports translation initiation.

Two other processes that also happen during transcript elongation are the removal of introns, and alternative splicing.[15] About 95% of human genes are thought to be alternatively spliced, resulting in about 100,000 different proteins being expressed from the estimated ~22,000 human genes.[16] The repertoire of possible splice variants for each gene is encoded in the DNA in the form of splice-site consensus sequences. These sites are bound on the nascent RNA by site-specific post-translationally modified proteins that enhance or repress splicing at the site in a combinatorial manner.[17] Mutations causing incorrect splicing of the β-globin transcript are a common cause of β-thalassemia (a severe anemia).[18]

- The transcript is completed in different ways for messenger, ribosomal, and transfer RNAs. For mRNA molecules, a motif of six nucleotides (AAUAAA) followed by a GU-rich sequence acts to recruit various factors to the nascent RNA. The factors dissociate the transcript from the transcription complex, and cap the RNA 3′ tail with hundreds of

---

[15] Examples of alternative splicing include: skipped exons, exons made longer or shorter, exons combined from adjacent genes via a read-through transcript, and alternative usage of 5′ or 3′ noncoding sequences.

[16] Q Pan *et al.*, Deep surveying of alternative splicing complexity in the human transcriptome by high-throughput sequencing, *Nature Genetics*, 2008 **40**(12): 1413–1415. M Sultan *et al.*, A global view of gene activity and alternative splicing by deep sequencing of the human transcriptome, *Science*, 2008, 321(5891): 956–960. ET Wang *et al.*, Alternative isoform regulation in human tissue transcriptomes, *Nature*, 2008, **456**(7221): 470–476. See also the Alternative Splicing and Transcript Diversity Database (ASTD): http://www.ebi.ac.uk/astd/. G Koscielny *et al.*, ASTD: The Alternative Splicing and Transcript Diversity database, *Genomics*, 2009, **93**: 213–220.

[17] M Soller, Pre-messenger RNA processing and its regulation: a genomic perspective, *Cell. Mol. Life Sci.*, 2006, **63**: 796–819. DL Black, Mechanisms of alternative pre-messenger RNA splicing, *Annual Review of Biochemistry*, 2003, **72**: 291–336.

[18] TJ Ley *et al.*, RNA processing errors in patients with β-thalassemia, *PNAS*, 1982, **79**(15): 4775–4779. GF Atweh, β-thalassemia resulting from a single nucleotide substitution in an acceptor splice site, *Nucleic Acids Research*, 1985, **13**(3): 777–790. In Cyprus, where one in seven people were found to be carriers, carrier screening has been performed for over 20 years (described in Chapter 6 of *Heredity and Hope* by Ruth Schwartz Cowan, 2008, Harvard University Press).

A bases (called poly(A)). More than half of all human transcripts have multiple 3′ termination sites, mostly in the several hundred untranslated bases preceding the poly(A) tail.[19] Alternative termination sites can affect mRNA stability, translation rate, and localization.[20] Mutations in the poly(A) tail itself can also cause diseases such as oculopharyngeal muscular dystrophy.[21]

- The sequence of transcribed RNA molecules is sometimes differentially modified in specific cells by a process known as RNA editing. In humans, the best known case is the editing of Apolipoprotein B (ApoB). ApoB is present in the blood in two isoforms. The liver synthesizes a full length isoform, whereas in the intestine, a shorter isoform is synthesized from the same transcribed RNA by the post-transcriptional replacement of a single nucleotide to create an alternate (early) stop codon.[22]

- The mature mRNA is transported to the cytoplasm, where its 5′ cap recruits the components of the ribosomal translation machinery. Once bound to the mRNA, the ribosomal complex scans down the mRNA for the translation start codon (5′-AUG-3′). During translation mRNA nucleotide triplets are matched to specific amino acids carried by transfer RNA molecules. Meanwhile, a sequence of protein–protein and protein RNA interactions results in the 3′ poly(A) tail of the mRNA also being recruited to the ribosomal complex.

Once the first ribosome has cleared the translation initiation site, a second ribosomal complex can be recruited to the mRNA. Because of the circularization of mRNA during translation, as soon as a ribosomal complex completes a round of protein synthesis, it can be immediately recruited by the 5′ cap of the mRNA for another round of translation. There are 5–10 times more ribosomes than mRNA molecules in a typical mammalian cell. In this way, a single mRNA molecule can be rapidly synthesized into many protein molecules.

Protein synthesis is energetically expensive. Moreover, erroneously synthesized proteins can behave in potentially damaging ways. A multitude of mechanisms insure against

---

[19] H Zhang, JY Lee and B Tian, Biased alternative polyadenylation in human tissues, *Genome Biology*, 2005, **6**(12): R100.

[20] J Hesketh, 3′-untranslated regions are important in mRNA localization and translation: lessons from selenium and metallothionein, *Biochemical Society Transactions*, 2004, **32**(6): 990–993.

[21] T Müller *et al.*, Genetic heterogeneity in 30 German patients with oculopharyngeal muscular dystrophy, *J. Neurology*, 2006, **253**(7): 892–895.

[22] DM Driscoll and E Casanova, Characterization of Apoliporotein B mRNA editing activity in Enterocyte extracts, *Journal of Biological Chemistry*, **265**(35): 21401–21403.

protein synthesis errors. mRNA molecules without 5′ or 3′ caps are rapidly degraded, and various "editing" and "proof-reading" mechanisms guard against incorrect amino acid incorporation during synthesis.[23] Nonetheless, a large variety of diseases ranging from cancer (melanoma, gastric) to infant-onset-diabetes and several neurological disorders are known to be due to malfunctions in protein synthesis.[24]

Condition-Specific Regulation of Gene Expression

As the examples in the preceding discussion illustrate, the regulation of gene expression in mammals can occur at every step of the transcription and translation processes. Here, we review some of the key regulatory mechanisms in order to provide a conceptual framework for later discussions of gene expression variability within and among individuals.

Condition-specific regulation of gene expression typically starts by (1) expression of constitutively active regulatory factors, and/or (2) post-translational activation of constitutively expressed regulatory factors. Transcription factor activation can be mediated by chemical modifications (e.g. phosphorylation), interactions with other proteins (e.g. production or degradation of a repressor protein), or spatial localization (e.g. transport of a regulatory factor to the nucleus). We will discuss the molecular interaction networks that underlie these upstream processes in the next section.

The key condition-specific mechanisms of gene expression include:

- **Regulation of chromatin (de)condensation.** We noted earlier that compacted chromatin is transcriptionally silent. The histone molecules in chromatin have long dangling tails (N and C terminal domains) which are subject to post-translational covalent modifications at multiple sites, including: acetylation, methylation, phosphorylation and ubiquitylation. Histone modifications regulate chromatin state in a combinatorial and context-dependent manner. In general (but not always), histone acetylation (via histone acetyl transferases, HATs) correlates with chromatin de-condensation and gene expression, whereas histone deacetylation (via histonedeacetylases, HDACs) results in condensation and transcriptional repression.

  Some histone modifications are highly gene-specific, while others affect large numbers of genes. Also, some histone methylation states are stable and can be passed on to

---

[23] L Cochella and R Green, Fidelity in protein synthesis, *Current Biology*, 2005, **15**(14): 536–540.
[24] GC Scheper, MS van der Knaap and CG Proud, Translation matters: protein synthesis defects in inherited disease, *Nature Reviews Genetics*, 2007, **8**: 711–732.

daughter cells. Each daughter cell receives half the parental histones. The inherited methylated histones then catalyze the methylation of nearby new histones.[25]

Histone methylation can be mediated by non-coding (regulatory) RNAs as well as various DNA-binding proteins and enzymes. Histone phosphorylation and ubiquitylation often occur in step with acetylation and methylation (respectively) and are thought to contribute to the positive (self-reinforcing) feedback loops that maintain histone modification states.

A number of additional mechanisms, such as replacement of histones with variants and histone destabilization (turn over) can further confer "epigenetic" (i.e. not coded in DNA sequence) regulation of gene expression.[26] Direct methylation of GC nucleotide pairs in DNA also silences transcription, probably by facilitating histone methylation. After cell division, specialized machinery methylates cytosine residues on the newly synthesized strand of DNA to mirror the methylation pattern on the parent strand. Surprisingly, DNA methylation is also used for transient repression of some genes.[27] Inappropriate epigenetic histone and DNA markers have so far been found to underlie more than a dozen diseases, including several cancers.[28]

- **Regulation of nucleosome positioning.** Nucelosomes can inhibit the formation and translocation of the basal transcription machinery at promoters. Before transcription initiation, DNA must be freed from the nucleosome complex. Many sequence-specific transcription factors (e.g. the glucocorticoid receptor) recruit nucleosome repositioning complexes.[29] In this way, nucleosome repositioning can provide an additional condition-specific mode of transcriptional regulation.
- **Regulation of the rate of transcription initiation.** In mammalian cells, transcription initiation is usually regulated by the binding of multiple, interacting, sequence-specific transcription factors to promoter and enhancer regions. Such factors typically bind cooperatively to short (six to 20 base-pair) sequence motifs in the major groove of the

---

[25] C Martin and Y Zhang, Mechanisms of epigenetic inheritance, *Current Opinion in Cell Biology*, 2007, **19**(3): 266–272. P Cheung and P Lau, Epigenetic regulation by histone methylation and histone variants, *Molecular Endocrinology*, 2005, **19**(3): 563–573.

[26] S Henikoff, Nucleosome destabilization in the epigenetic regulation of gene expression, *Nature Reviews Genetics*, 2008, **9**(1): 15–26.

[27] S Kangaspeska *et al.*, Transient cyclical methylation of promoter DNA, *Nature*, 2008, **452**: 112–115.

[28] DB Seligson *et al.*, Global histone modi?cation patterns predict risk of prostate cancer recurrence, *Nature*, 2005, **435**(7046): 1262–1266. For a review of epigenetic diseases and treatments, see G Egger *et al.*, Epigenetics in human disease and prospects for epigenetic therapy, *Nature*, 2004, **429**: 457–463.

[29] J Mellor, The dynamics of chromatin remodeling at promoters, *Molecular Cell*, 2005, **19**: 147–157.

DNA double helix. Multiple protein–protein and protein–DNA interactions bend the DNA such that the enhancer complex localizes to the promoter neighborhood[30] and recruits and activates the transcription apparatus.

Combinatorial regulation of transcription initiation by transcription factors allows precise control of gene expression. A typical human gene is regulated by dozens of condition-specific transcription factors. Groups of interacting factors cluster within DNA regions dubbed *cis*-regulatory modules (CRMs).[31] Each CRM typically spans several hundred base pairs and quantitatively regulates gene expression under specific conditions. A given enhancer region may contain multiple CRMs. Figure 2.2 shows an example mouse gene (*Sfpi1*, which encodes the PU.1 protein) and the binding sites of some of the factors known to regulate it.[32]

In the figure, the thick horizontal lines denote DNA. The right-angle arrow indicates the transcription start site (TSS). The black boxes indicate the exons (the spaces in

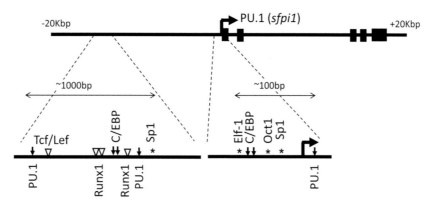

Figure 2.2: Transcription factor binding sites in two regulatory regions associated with *Sfpi1*.

---

[30] See for example IS Mastrangelo *et al.*, DNA looping and Sp1 multimer links: A mechanism for transcriptional synergism and enhancement, *PNAS*, 1991, **88**: 5670–5674.

[31] For a detailed discussion of the organizational principles underlying transcriptional regulation and gene regulatory networks, see EH Davidson, *The Regulatory Genome: Gene Regulatory Networks in Development and Evolution*, Academic Press, 2006.

[32] Figure based on M Hoogenkamp *et al.*, The PU.1 locus is differentially regulated at the level of chromatin structure and noncoding transcription by alternate mechanisms at distinct developmental stages of hematopoiesis, *Molecular and Cellular Biology*, 2007, **27**(21): 7425–7438.

between are the introns). The exons span a region of about 19 Kbp downstream of the TSS. The insets show some of the known transcription factor binding sites within the proximal promoter, and a distal enhancer about 14 Kbp upstream of the TSS. The sites marked with downward arrows are primarily active in myeloid cells, whereas the sites marked with triangles are primarily active in T-cell development. The indicated sites mediate only a small subset of the regulatory influences on *Sfpi1*. Correct PU.1 expression in transgenic mice requires a 91 Kbp fragment of genomic DNA.[32]

The above example was for a single gene. In any given cell at any given time, large numbers of transcription factors cross-regulate each other, resulting in complex gene regulatory networks (GRNs) that determine cellular state and regulate processes such as embryonic development. Mammalian GRNs that regulate specific cellular processes typically span dozens of genes.[33] As an illustrative example, figure 2.3 shows a small portion of the GRN thought to underlie early T cell development.[34]

In these figures, genes are indicated by symbols similar to the preceding figure (horizontal lines and right-angle arrows). Lines emanating from each gene indicate the

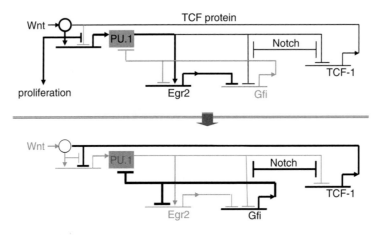

Figure 2.3: A gene regulatory network in successive developmental states.

---

[33] For interactive time-course visualizations of example mammalian developmental gene regulatory networks, see http://www.its.caltech.edu/~tcellgrn/TCellMap.html and http://www.mcb.harvard.edu/McMahon/BioTapestry.

[34] This figure is based on perturbation data, so some of the interactions may be indirect. Figure based on EV Rothenberg, Regulatory factors for initial T lymphocyte lineage specification, *Current Opinion in Hematology*, 2007, **14**(4): 322–329.

regulatory targets of the gene's protein product. A line ending in an arrowhead indicates a positive regulatory influence, while a line ending in a bar indicates a repressive role. The disk at the top left indicates protein–protein interactions between components of the Wnt signaling system and the TCF-1 transcription factor. In the absence of Wnt signaling TCF-1 binds its target with a co-repressor and inhibits transcription, while in the presence of Wnt signaling, TCF-1 acts as a transcriptional co-activator.

The top panel shows the putative state of this small GRN fragment in early T cell precursors shortly after the cells arrive in the thymus. The lower panel shows the state of the same network a few days later. The *Sfpi1* gene (encoding the PU.1 protein) is at the top left. Early on, its expression is supported by Wnt signaling. Later, in the absence of Wnt signaling (gray lines indicate inactive/absent regulatory interactions), *Sfpi1* is turned off. *Gfi1*, which is repressed by PU.1 in early T cell precursors, becomes active once PU.1 is repressed, and in turn represses *sfpi*/PU.1 activity.

We see that the two developmental states illustrated differ by mutually exclusive expression patterns of *Sfpi1*/PU.1 and its targets (*Gfi1* and *Egr2*). Note also the complementary and coordinated role of Notch signaling, which intensifies in the lower panel and counteracts any inhibitory action by remaining PU.1 protein molecules on *TCF-1*.

We noted earlier that regulation of chromatin state also affects transcription. The cause and effect relationship between direct transcriptional regulation, and regulation of chromatin is complex. In yeast, it has been suggested that chromatin state acts as an all-or-none (permissive/repressive) expression switch, while site-specific regulatory factors regulate the level of gene expression[35] as a function of the concentrations of the regulatory transcription factors.[36]

In addition to enhancer-mediated recruitment of the basal transcription apparatus, a variety of molecular mechanisms are used to regulate the rate at which new transcripts are initiated. For example, in some genes, the basal transcription apparatus is pre-formed, but stalled near the start site. Activating factors then release the RNA polymerase from its paused state to initiate transcription.[37] Apart from releasing paused transcription

---

[35] FH Lam, DJ Steger and EK O'Shea, Chromatin decouples promoter threshold from dynamic range, *Nature*, 2008, **453**: 246–250.
[36] See for example, H Bolouri and EH Davidson, Modeling transcriptional regulatory networks, *BioEssays*, 2002, **24**: 1118–1129.
[37] LJ Core and LT Lis, Transcriptional regulation through promoter proximal pausing of RNA polymerase II, *Science*, 2008, **319**: 1791–1792.

complexes, transcription elongation factors may also directly regulate the *rate* of transcript elongation.[38]

- **Differential RNA splicing and editing.** Enhancer and silencer sites on transcribed RNA are bound by regulatory proteins, which act individually or in combination to determine the splicing of nascent RNAs. In this way, condition-specific protein–protein reactions (e.g. due to a signaling event) can determine the splice-type of genes undergoing transcription. Between 15% and 50% of disease-associated exonic mutations are thought to involve inappropriate splicing.[39]

  We mentioned RNA editing in the previous section. The predominant form of RNA editing in mammals is adenosine-to-inosine (A-to-I) editing. Inosine acts as guanosine during translation. Thus, A-to-I conversion in coding sequences leads to an amino acid change and can therefore affect protein function. A-to-I editing is catalyzed by enzymes called ADARs (adenosine deaminases acting on RNAs). ADARs are particularly active in the brain, exhibiting temporal changes in activity as well as changes in sub-cellular localization. In particular, differential editing of neural RNAs appears to play a key role in brain development and learning.[40]

- **Regulation of mRNA transport.** mRNA export from the nucleus requires export factors that bind to protein complexes formed at exon junctions (splicing sites). mRNAs without exon junctions are either not exported or recruit specialized export proteins. Many viruses (e.g. herpes and HIV-1) require the export of intron-less RNAs for proliferation. The genomes of these viruses encode regulatory RNAs which recruit and activate specific nuclear export proteins.[41]

- **Regulation of translation.** A common mechanism for condition-specific regulation of translation is RNA interference, in which small RNA molecules typically bind complementary mRNA molecules and mediate their degradation or block their translation.[42] Dysfunctional RNA interference underlies a variety of human diseases.

---

[38] M Endoh *et al.*, Human Spt6 stimulates transcription elongation by RNA polymerase II *in vitro*, *Molecular and Cellular Biology*, 2004, **24**(8): 3324–3336.

[39] L Cartegni, SL Chew and AR Krainer, Listening to silence and understanding nonsense: exonic mutations that affect splicing, *National Reviews Genetics*, 2002, **3**: 285–298.

[40] JS Mattick and MF Mehler, RNA editing, DNA recoding and the evolution of human cognition, *Trends in Neurosciences*, 2008, **31**(5): 227–233. S Maas *et al.*, A-to-I RNA editing and human disease, *RNA Biology*, 2006, **3**(1): 1–9.

[41] RM Sandri-Goldin, Viral regulation of RNA export, *Journal of Virology*, 2004, **78**(9): 4389–4396.

[42] H Großhans and W Filipowicz, The expanding world of small RNAs, *Nature*, 2008, **451**: 414–416.

Regulatory RNAs appear to play a key role in the development and function of immune cells[43] and neurons. In Alzheimer's disease (AD), an inappropriately expressed enzyme (β-secretase) cleaves the amyloid precursor protein into its neuro-toxic form. Two regulatory RNAs can act upon β-secretase. One operates in healthy cells and degrades the β-secretase mRNA. The second regulatory RNA is activated by β-amyloid particles and counteracts the former in AD, creating a vicious circle.[44]

Inappropriate RNA interference is also present in many cancers because the tumor suppressor protein p53 activates the transcription of a family of regulatory RNAs.[45] Another example of condition-specific regulation of translation occurs in response to viral attacks. The immune cells respond by synthesizing and secreting interferon-family cytokines, which in turn inhibit (viral) mRNA translation.[46]

- **Post-translational regulation of protein activity.** A remarkable array of chemical modifications can regulate the function of synthesized proteins.[47] We already noted the post-translational acetylation, methylation, ubiquitylation and phosphorylation of histone proteins in chromatin state regulation. Other examples of post-translational chemical modification include glycosylation, sumoylation and biotinylation.

The diversity of chemical states that proteins can adopt vastly increases their functional repertoire. Chains of chemical reactions can amplify even small biochemical differences between protein states, giving each modified form a highly specific function.[48] Moreover,

---

[43] RM O'Connell *et al.*, Sustained expression of microRNA-155 in hematopoietic stem cells causes a myeloproliferative disorder, *Journal of Experimental Medicine*, 2008, **205**(3): 585–594; D Baltimore *et al.*, MicroRNAs: new regulators of immune cell development and function, *Nature Immunology*, 2008, 9(8): 839–845.

[44] MA Faghihi *et al.*, Expression of a noncoding RNA is elevated in Alzheimer's disease and drives rapid feed-forward regulation of b-secretase, *Nature Medicine*, 2008, **14**(7): 723–730; SS Hebert and B De Strooper, Alterations of the microRNA network cause neurodegenerative disease, *Trends in Neurosciences*, 2009, **32**(4): 199–306.

[45] L He *et al.*, A microRNA component of the p53 tumour suppressor network, *Nature*, 2007, **447**(7148): 1130–1134.

[46] Interferon repression of translation may occur through multiple mechanisms; see Y He *et al.*, Regulation of mRNA translation and cellular signaling by Hepatitis C virus nonstructural protein NS5A, *Journal of Virology*, 2001, **75**(11): 5090–5098; MZ Tesfay *et al.*, Alpha/Beta interferon inhibits cap-dependent translation of viral but not cellular mRNA by a PKR-independent mechanism, *Journal of Virology*, 2008, **82**(6): 2620–2630.

[47] For a comprehensive review see CT Walsh, *Posttranslational Modification of Proteins: Expanding Nature's Inventory*, Roberts and Company, 2006.

[48] PS Swain and ED Siggia, The role of proofreading in signal transduction specificity, *Biophysical Journal*, 2002, **82**: 2928–2933; OJG Somsen *et al.*, Selectivity in overlapping MAP Kinase cascades, *Journal of Theoretical Biology*, 2002, **218**: 343–354.

proteins in modified states provide the cell with a form of short-term memory of past events, enabling physiological features such as priming (increased sensitivity) and tolerance (reduced sensitivity). The combination of the longer-term memory provided by chromatin and DNA modifications with the short-term memory provided by protein modifications enables cells to functions as complex information processing systems on multiple timescales.

Modified proteins differentially regulate cellular processes in a variety of ways, including rapid degradation, buffering and transport between organelles, complex formation, conformational changes, and cleavage. In this way, thousands of cellular processes can progress simultaneously without adversely affecting each other. In practice most cellular processes are tightly coordinated with one another through cross-regulatory links and feedback mechanisms.

We will discuss the coordination of physiological and developmental processes in the next section. We end this section by noting that the basal transcription complex includes and recruits enzymes that repair DNA damage during transcription. The tight coupling between transcription and DNA maintenance is highlighted in Xeroderma Pigmentosum (extreme UV sensitivity), where mutations in the core transcription factor TFIIH result in defective repair of UV-damaged DNA.[49]

Molecular Interaction Networks and Systems

Cellular processes are traditionally presented as linear chains of events. For example, signaling pathways are typically described as involving a sequence of events such as:

1. Ligand binds receptor.
2. The bound receptor modifies (activates) an effector molecule (directly, or through a chain of reactions, as in the MAP Kinase cascade).
3. One or more transcription factors are modified by the activated effector molecules.
4. Activated factors bind DNA and initiate transcription of downstream genes.

Such simplification is often *necessary* in order to convey a succinct view of key events during signal transduction. But as a representation of how cellular processes are regulated, it is

---

[49] F Coin *et al.*, Mutations in XPB and XPD helicases found in xeroderma pigmentosum patients impair the transcription function of TFIIH, *EMBO Journal*, 1999, **18**(5): 1357–1366.

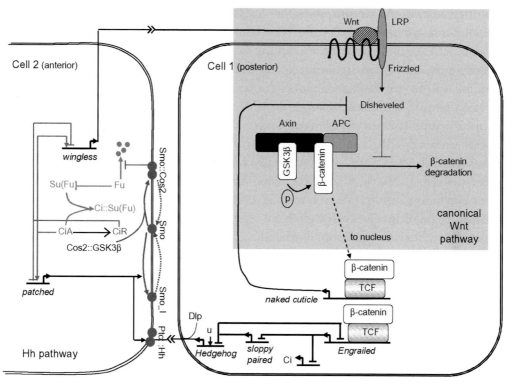

Figure 2.4: Canonical Wnt signaling and its multicellular context during fruit fly development.

highly misleading. Figure 2.4[50] presents two views of the Wnt signaling pathway. The portion on gray background shows the canonical Wnt signal transduction pathway as it is usually presented (each blob represents a molecular species; right-angle arrows represent genes, straight arrows represent activation or transition, and lines ending in a bar denote repression).

The canonical Wnt signaling pathway is involved in multiple cancers, and more than half a dozen other genetic diseases.[51] Briefly, in the absence of Wnt signaling, Disheveled molecules are part of a complex of proteins that together degrade β-catenin in the cytoplasm. Wnt ligands bind Frizzled-LRP receptor complexes, resulting in phosphorylation of Disheveled, and disruption of

[50] I am grateful to Drs Alfonso Martinez-Arias (Cambridge University, UK) and Randy Moon (University of Washington, USA) for advice on this figure.
[51] CY Logan and R Nusse, The Wnt signaling pathway in development and disease, *Annual Reviews Cell and Developmental Biology*, 2004, **20**: 781–810.

β-catenin degradation. As a result, β-catenin levels build up in the nucleus. Nuclear β-catenin dislodges the transcriptional co-repressor Groucho (not shown) and binds to the transcription factors TCF/LEF, resulting in the transcription of TCF/LEF target genes.

In the figure above, we see how the canonical Wnt signaling pathway (on gray background) fits within an example cellular context. Since molecular interactions are better understood in model organisms than in humans, this example relates to an early specification process in Drosophila embryos. Two cells are shown. In cell 1, the *Engrailed* gene is activated by an earlier developmental process and represses *Sloppy paired*.

Engrailed and *Sloppy paired* are mutually repressive. In cell 2, *Sloppy paired* represses *Engrailed*. In cell 1, *Sloppy paired* repression by *Engrailed* results in de-repression of the *Hedgehog* gene. As a result, cell 1 transmits a Hedgehog (Hh) signal to cell 2. Hh signal transduction in cell 2 involves a complex set of protein–protein interactions involving multiple feedbacks. These interactions ultimately activate transcription of the *Wingless* gene encoding Wnt, and secretion of the Wnt ligand. Wnt signal transduction in cell 1 maintains *Engrailed* expression, thus setting up a self-sustaining inter-cellular positive feedback loop.

The larger interaction network represented in the above figure has a number of functional characteristics that are not apparent in the representation of the canonical Wnt signaling pathway (gray box). For example, this system of interactions behaves as a highly stable, noise-resistant bistable (on/off) switch, ensuring stereotyped developmental patterns irrespective of variations in cellular content, size, temperature, etc. This robustness is achieved through multiple feedback interactions[52] that couple Wnt signal transduction in cell 1 to Hh signaling in cell 2.

The myriad molecular species that perform the house-keeping and physiological functions of cells interact widely with each other. These interactions tightly coordinate the many biochemical processes that underlie cellular physiology. They also make it very difficult to predict the effect of a structural, functional, or quantitative change in any component molecule on all aspects of cellular physiology. In the above example network, mutations that change the degradation rate of any protein by a factor of 10 have little effect on the pattern forming capabilities of the system.[53] On the other hand, a mutation in the Hh receptor *Patched* (Ptc) that disrupts

---

[52] Some example feedback loops in the interactions shown are: (*Nkd* gene and Disheveled protein), (*En* and *Slp* genes), (*Ptc* and *Wnt* genes), the regulation of the chemical state of CI into activating (CIA) and repressive (CIR) forms by multiple interactions among Ptc, Smo, Cos2, Fu, and Su(Fu), and finally the positive inter-cellular feedback loop between Wnt and Hh signaling pathways.

[53] See G. von Dassow *et al.*, The segment polarity network is a robust developmental module, *Nature* 2000, **406**: 188–192; NT Ingolia, Topology and robustness in the Drosophila segment polarity network, *PLoS Biology*, 2004, **2**(6): 805–815.

Ptc:Hh dimerization will demolish the model's pattern formation capability (because it disrupts the transmission of the initial embryonic pre-pattern between the cells).

The multi-cellular, multi-pathway network presented in the above figure is itself a highly simplified representation that does not show many additional regulatory interactions. For example, in both fruit flies and mice, Notch signaling modulates the effect of Wnt signaling[54] (which may explain the tumor suppressor characteristic of Notch). Thus, a mutation in a particular element in a particular pathway may have effects on many other functionally distinct pathways and processes. A further complication is that the same gene/protein typically takes part in many distinct cellular processes in different cells and conditions. For example, GSK3β, a component of the canonical Wnt signaling pathway, also takes part in glucose metabolism in muscle cells, and in axon guidance in developing neurons. In each case, GSK3β acts in a different pathway (with its own cohort of associated cellular processes). Dysregulated GSK3β activity has been associated with cancer,[55] Alzheimer's disease[56] and diabetes.[57] But in each case, the mechanisms of dysregulation, and the pathways affected, are different.

We will use Wnt signaling and its components as a recurring example pathway in the rest of this book.

Cross-Regulation and Physiology at the Organ Level

The preceding discussions emphasized regulatory processes and interdependencies at the molecular level. Considerable regulation and cross-regulation also occurs at the organ level. In the context of this book, this is an important consideration because it emphasizes the need to relate diagnoses to both dysregulated intra-cellular pathways, and also to organ physiology. This section briefly discusses interactions among some key organs and the implications for health maintenance, the processing of drugs, and disease detection (see also Chapter 5).

Collective changes in cellular state drive organ function. The state of one organ can influence the behavior and function of cells in other organs via hormonal, neuronal, immune and other signaling mechanisms. For example, extra-cellular, intra-cellular, and blood (plasma) dehydration can arise from a variety of shared and distinct causes. Dehydration signals

---

[54] P Hayward *et al.*, Notch modulates Wnt signalling by associating with Armadillo/β-catenin and regulating its transcriptional activity, *Development*, 2005, **132**: 1819–1830.

[55] P Polakis, Wnt signaling and cancer, *Genes and Development*, 2000, **14**: 1837–1851.

[56] W Noble *et al.*, Inhibition of Glycogen Synthase Kinase-3 by lithium correlates with reduced tauopathy and degeneration *in vivo*, *Proc. Nat. Acad. Sci. USA* 2005, **102**(19): 6990–6995.

[57] EJ Henriksen, Modulation of muscle insulin resistance by selective inhibition of GSK3 in Zucker diabetic fatty rats, *Am. J. Physiol. Endocrinol. Metab.* 2003, **284**: E892–E900.

(e.g. vasopressin, angiotensin) from the gastrointenstinal (GI) tract, the liver, the kidneys, and the circulatory system are integrated in the brain, which may then generate a desire for water, salty foods, etc. Moreover, between the sensing of dehydration and its resolution, organ physiology is modified to minimize the adverse effects, for example by reabsorbing sodium and water from urine, and by redistributing blood and interstitial fluids.[58]

In this context, it is useful to think of blood as an organ that is distributed throughout the body. Blood offers a convenient means of monitoring the entire body and so will be important to our discussion of molecular biomarkers in Chapter 7. A 150-pound (68 kg) person has about 4.7 liters of blood. Just under half of this volume is taken up by red blood cells (oxygen transporters), white blood cells (various immune cells), and platelets (coagulating cells). The vast majority (about 99%) of cells in blood are red blood cells (erythrocytes). Erythrocytes and platelets do not have nuclei. Thus, when blood is used for DNA sequencing, it is really the white blood cells (leukocytes) that are sequenced.

There are about 5,000–10,000 immune cells (leukocytes) in a 1 mm^3 drop of blood. Immune cells act as sentinels against diseases occurring anywhere in the body. They also change state during the course of an illness. Thus, assaying the state of blood-borne immune cells offers an attractive strategy for detecting and monitoring diseases[59] (see Chapters 4 and 7).

The liquid part of blood (about 55% of total) is called plasma. Blood serum is plasma stripped of clotting factors. Blood delivers oxygen and nutrients and takes away waste products — including fragments of dead cells — from all tissues. It also acts as a conduit for the distribution of secreted signaling proteins such as cytokines. For this reason, blood serum is a rich source of health status indicators (biomarkers), as will be discussed in Chapter 7.

The composition of blood is jointly — and tightly — regulated by the liver and the kidneys. The liver is responsible for the uptake of nutrients, glucose storage, the synthesis of bulk serum proteins, and the modification of drugs and other xenobiotics into forms that can be filtered by the kidneys. In addition to filtering waste products, the kidneys regulate blood volume, pressure and the rate at which new red blood cells are produced. On the other hand, aging and damaged red blood cells are removed from the blood stream in the liver and the spleen.

The concentration of erythrocytes in the blood is an example of multi-organ homeostatic regulation. It can be visualized as the regulation of water flow into and out of a water reservoir,

---

[58] MJ McKinley and AJ Johnson, The physiological regulation of thirst and fluid intake, *News in Physiological Sciences*, 2004, **19**: 1–6.

[59] See for example, KS Anderson and J LaBaer, The sentinel within: exploiting the immune system for cancer biomarkers, *Journal of Proteome Research*, 2005, **4**(4): 1123–1133; P Duramad *et al.*, Flow cytometric detection of intracellular th1/th2 cytokines using whole blood: validation of immunologic biomarker for use in epidemiologic studies, *Cancer Epidemiology, Biomarkers and Prevention*, 2004, **13**(9): 1452–1458.

such that the water level is held constant. This type of active maintenance of a state at a given set-point (homeostasis) is usually achieved via negative feedback loops that counteract change. One upshot of such regulation is that small perturbations to the system are compensated for and may produce no external symptoms. However, beyond a certain threshold (analogous to fully opening or shutting the flood gates), perturbations quickly become symptomatic.

Tight regulatory coupling is common among the organs of the body. In the discussions in subsequent chapters, readers may find the above "water reservoir" analogy helpful in thinking about the effects of mutations in genes, responses to drugs or toxic materials, and other perturbations to the human body.

The Genome in its Environmental Context

The brief review presented in this chapter has highlighted two important, interacting, and co-dependent forces shaping human form and function. Firstly, information encoded in our DNA defines the repertoire of the molecular capabilities of cells. This is the basis for

Table 2.1: Mechanisms of differential genome interpretation at various scales.

Point of action	Causes of variation
DNA	Methylation Histone modification
RNA	Editing Degradation Alternative Splicing Chemical modification
Protein	Degradation Chemical modification Localization Complex formation
Cell	Differentiation Cell-cell interactions Proliferation rate Apoptosis
Organ	Exercise, diet, drugs, aging, hormones, stress

the role of DNA sequence variations in generating human diversity, as discussed in the next chapter.

The second take away message from this chapter is that the translation of genomically encoded information into cellular function is frequently dependent on extra-cellular conditions. In this way, the environment can modulate cellular behavior both in the short term and also in the long term, as will be discussed in Chapter 4.

Table 2.1 summarizes some of the key mechanisms through which environmental influences can modulate the interpretation of genomic information at the molecular level.

We noted earlier that feedback loops are prevalent within cells, between cells, and between organs. Feedbacks also coordinate physiological processes across scales: upwards from cells to organs to organism, and downwards from organism to organs to cells.

Feedbacks enable the body to maintain operating conditions such as temperature and PH, but they also make it difficult to predict the response of the body to novel perturbations such as previously uncharacterized gene mutations and drug treatments. These issues will be discussed further in Chapter 5.

CHAPTER 3

DNA Sequence Variations, their Prevalence, and Effects on Cellular Biochemistry

This chapter reviews current understanding of the mechanisms, types and abundance of DNA sequence variations among individuals. The likely biochemical outcomes of each type of sequence variation are discussed. Our focus in this chapter is primarily on germline (heritable) variation. Somatic mutations and modifications will be discussed in the next chapter.

The overview presented here is necessarily highly abbreviated. Genetics is a discipline rich in tradition and complex in terminology. In particular, DNA sequence variants are often referred to by different names depending on context and emphasis.[1] To avoid confusion and to emphasize concepts of special relevance to this book, we have avoided jargon as far as possible. For more information please refer to standard medical genetics and molecular biology textbooks[2] as well as the publications cited herein.

Sources of DNA Sequence Variations

DNA sequence variations can arise in both the germline cells (whose DNA will be passed to offspring) and also in somatic cells (the majority of cells in the human body, e.g. liver, heart, muscle, neuron, etc.).

Recall that germline cells undergo meiosis to produce (haploid) gametes (sperm and egg cells) and that crossover of homologous chromosomes during recombination in effect shuffles

---

[1] SW Scherer *et al.*, Challenges and standards in integrating surveys of structural variation, *Nature Genetics Supplement*, 2007, **39**: S7–S15.

[2] See for example JD Watson *et al.*, *Molecular Biology of the Gene*, 6th Edition, Cold Spring Harbor Press, 2008; RL Nussbaum, RR McInnes and FW Willard, *Thompson and Thompson Genetics in Medicine*, 6th Edition, Elsevier, 2007.

the parental chromosomes in germ cells. Aside from possible replication errors, meiosis introduces two types of variability into the genomes of offspring. Firstly, an offspring will inherit only one of each parent's 23 pairs of chromosomes. Thus, each parent's genome can generate 2^{23} (i.e. over 8 million) distinct gametes. For any given pair of parents, the probability that two offspring will have exactly the same complement of chromosomes is one in 64 trillion.

A second source of variability arising during meiosis is crossover. Each chromosome is estimated to undergo an average of two to three crossovers per meiosis.[3] Crossovers occur at sites where double-stranded DNA is cut open by an enzyme activated during meiosis (SPO11). The cuts are repaired by subsequent recombination between homologous sequences on parental chromosomes. In this way, crossovers shuffle the genome among homologous chromosomes. SPO11 cuts DNA in a non-uniform manner, favoring nucleosome-free sequences such as promoters. Moreover, recombination events are more common at repetitive DNA sequences. The overall result is that crossover frequency varies considerably across chromosomes.[4]

Sometimes meiotic crossovers occur at unequal loci, leading to duplications and deletions in the resulting gametes, as illustrated in figure 3.1.

We will return to the topic of copy number variations later in this chapter. For now, we note that over evolutionary timescales, distant loci on a chromosome are more likely to be shuffled onto different chromosomes than nearby loci. Because the human species is relatively new, we may expect to find many chromosomal neighborhoods uninterrupted by crossover events. Such regions will therefore be near-identical among people with similar genetic background (i.e. people sharing a common ancestral history). Put another way, we can use any set of characteristic features in such regions (called haplotypes) to group people into distinct genetic populations. This concept has important implications for genetic testing, as will be discussed in Chapter 6.

In addition to recombination, a number of other mechanisms can introduce random shuffling, copying, deletion and inversion of chromosomal regions. In B and T-cells of the immune system, double-stranded DNA breaks are generated as part of the process of receptor diversity

---

[3] B Alberts *et al.*, *Molecular Biology of the Cell*, 4th Edition, Garland Press, 2002, p. 1017. Note that crossovers can also occur during somatic cell divisions (mitosis) if DNA is damaged (e.g. due to radiation).

[4] S Myers *et al.*, A fine-scale map of recombination rates and hotspots across the human genome, *Science*, 2005, 310(5746): 321–324; E Mancera *et al.*, High-resolution mapping of meiotic crossovers and non-crossovers in yeast, *Nature*, 2008, **454**: 479–485.

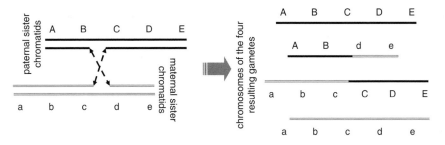

Figure 3.1:   How duplications and deletions can arise during crossovers.

generation (mediated by the *RAG1/2* genes). Since such genomic sequence variations are not heritable, we will discuss them in the next chapter.

Double-stranded DNA breaks also occur due to pathological causes such as ionizing radiation and free radicals. Such breaks are repaired by multiple pathways that rejoin the cut ends.[5] Errors in re-joining double-stranded DNA breaks are common and can lead to sequence deletions and duplications, as well as chromosomal instabilities characteristic of tumor cells.[6]

Dual break points in a single chromosome can lead to two additional forms of chromosomal rearrangement: segmental inversions and deletions, as illustrated in figure 3.2.

Two additional mechanisms can generate DNA variability in both germline and somatic cells: DNA rearrangements due to transposable elements, and DNA maintenance and replication errors.

Replication errors arise in part because — in spite of its multiple error avoidance and correction processes — DNA replication can never be perfect. They also arise because many environmental factors can modify DNA sequences. Example factors include chemical mutagens such as base analogs and DNA-intercalating aromatic compounds, and of course radiation. The effects of mutagens on DNA are varied. For example, base analogs generally substitute one nucleotide for another, while intercalating agents cause a frame-shift in the

---

[5] MR Lieber, The mechanism of human nonhomologous DNA end joining, *Journal of Biological Chemistry*, 2008, **283**(1): 1–5; JK Moore and JE Haber, Cell cycle and genetic requirements of two pathways of nonhomologous end-joining repair of double-strand breaks in Saccharomyces cerevisiae, *Molecular and Cellular Biology*, 1996, **16**(5): 2164–2173.

[6] S Espejel *et al.*, Mammalian Ku86 mediates chromosomal fusions and apoptosis caused by critically short telomeres, *EMBO Journal*, 2002, **21**(9): 2207–2219.

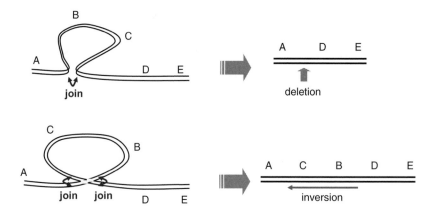

Figure 3.2:   How inversions and deletions can occur within individual chromosomes.

mapping of nucleotides to amino acids. Radiation, on the other hand, tends to increase the rate of crossovers and chromosomal end-fusions.

Ironically, water, which is necessary for maintenance of the DNA double helix structure, can also damage DNA. In particular, water-mediated de-amination of methylated Cs results in C to T transitions and loss of the transcriptional silencing character of methylated cytosines.[7] In thermophyllic bacteria this process can be regulated and therefore highly non-random.[8] But such regulation has so far not been reported in humans.

Finally, transposable elements (TEs) — thought to be (mostly) inactive remnants of viral DNA — cover nearly half the human genome and impact about 25% of promoters.[9] Over 90% of human TEs are so called retrotransposons. Retrotransposons first copy themselves to RNA and then back to DNA via a reverse transcriptase. They range in length from a few hundred to a few thousand nucleotides. Many TEs have been co-opted by evolutionary processes to sculpt transcriptional enhancers and promoters, or modified to act as new regulatory RNAs.[9,10]

---

[7] J-C Shen, WM.Rideout and PA Jones, The rate of hydrolytic deamination of 5-methylcytosine in double-stranded DNA, *Nucleic Acids Research*, 1994, **22**(6): 972–976.

[8] M Carpenter *et al.*, Sequence-dependent enhancement of hydrolytic deamination of cytosines in DNA by the restriction enzyme PspGI, *Nucleic Acids Research*, 2006, **34**(13): 3762–3770.

[9] C Feschotte, Transposable elements and the evolution of regulatory networks, *Nature Reviews Genetics*, 2008, **99**: 397–405.

[10] GJ Faulkner *et al.*, The regulated retrotransposon transcriptome of mammalian cells, *Nature Genetics*, 2009, **41**(5): 563–571.

Because they are widely repeated throughout the genome, TEs also act as hot spots for recombination (chromosomal crossover). TE-related mutations are known to underlie over a dozen genetic diseases, and many cancers.[11] TEs are also involved in epigenetic programming of the genome, as will be discussed in Chapter 4.

Observed Types of DNA Sequence Variation

DNA sequence variations range in size from single nucleotides to whole chromosomes. Irrespective of the size, DNA sequence variations can be classified as single or multiple base substitutions, deletions, insertions, duplications (or, in general, amplifications), inversions, and re-locations (translocations). These are summarized schematically in figure 3.3 (colored boxes represent sequence features).

Single-base changes (also called point mutations) involve the replacement of one nucleotide by another — usually during DNA replication. They are the most frequent type of variation in the human genome. DNA loci at which repeated point mutations are observed are referred to as Single Nucleotide Polymorphisms (SNPs).

Point mutations in coding regions can change one amino acid to another. For example, in sickle cell anemia, replacement of A by T at the 17th nucleotide of the gene for the β chain of hemoglobin changes the codon GAG (glutamic acid) to GTG (valine).

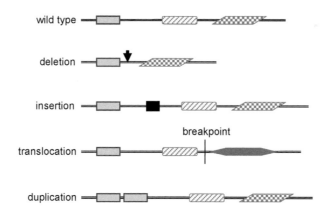

Figure 3.3: Some observed types of DNA sequence variation.

[11] LS Collier and DA Largaespada, Transposable elements and the dynamic somatic genome, *Genome Biology*, 2007, **8**(Suppl 1): S5; MA Batzer and PL Deininger, Alu repeats and human genomic diversity, *Nature Reviews Genetics*, 2002, **3**: 370–380.

If a point mutation changes an amino acid codon to one of the stop codons (TAA, TAG, or TGA), the translation of the messenger RNA will be terminated prematurely. The more truncated the protein product, the more likely it is to be dysfunctional. In cystic fibrosis, more than 1,500 mutations have been reported in a single gene that encodes a protein of 1,480 amino acids, called the *cystic fibrosis transmembrane conductance regulator* (*CFTR*). About 10% of these mutations produce premature stop codons.

Some point mutations change the codon triplet without changing the amino acid specified. Such "silent" mutations are assumed to have no phenotypic effect. However, a recent study suggests this may not always be the case.[12]

Splice-site mutations. Nucleotide motifs at splice sites guide the RNA splicing machinery. Mutations that alter splice sites can therefore result in splicing failures (e.g. skipped exons, or introns that are not spliced out). In addition, partial modification of a splicing enhancer motif, and insertion/deletion events that change the location of the motif, can cause regulatory failures in alternative splicing.[13] Splicing-related mutations are thought to be one of the most common disease-causing mutations in the coding region.[14]

Insertions and deletions can range in size from single nucleotides to entire chromosomes. Shorter (<100 base pairs) insertions and deletions are often referred to as "indels". In coding regions, indels whose length is not a multiple of three cause a "frameshift" in the nucleotide to amino acid translation. The resulting protein will have an altered amino acid sequence from that point onwards and so may be very different from the wild type protein. Frameshifts can also create new stop codons, resulting in premature termination of the protein sequence.

Indels of three nucleotides or multiples of three preserve the reading frame in coding sequences. However, a number of inherited human disorders are caused by the insertion of many copies of the same triplet of nucleotides. Huntington's disease and the fragile X syndrome are examples of such trinucleotide repeat diseases. In Huntington's disease, the repeated trinucleotide is CAG (glutamine). Unaffected individuals have between 11 and 34 repeats of CAG in the huntingtin protein (htt). Affected individuals have more than 37 CAG

---

[12] C Kimchi-Sarfaty *et al.*, A "silent" polymorphism in the MDR1 gene changes substrate specificity, *Science*, 2007, **315**(5811): 525–528.

[13] I Vorechovsky, Aberrant 3′ splice sites in human disease genes: mutation pattern, nucleotide structure and comparison of computational tools that predict their utilization, *Nucleic Acids Research*, 2006, **34**: 4630–4641.

[14] N Lopez-Bigas *et al.*, Are splicing mutations the most frequent cause of hereditary disease? *FEBS Letters*, 2005, **579**(9): 1900–1903.

repeats. Mutant htt protein has multiple toxic effects in neurons. In particular, it increases p53 activity leading to apoptosis.[15]

Duplications of larger DNA segments (100 b to 1 Mb) are estimated to cover about 150 Mb of the human genome, and are non-uniformly distributed.[16] Current data suggests that the length distribution of multiply copied DNA segments is approximately exponential, with many smaller and fewer larger variants.[17] Diseases in which large portions of chromosomes have been deleted, duplicated, or rearranged form a large family collectively known as "genomic disorders".[18]

Most insertions and deletions of whole chromosomes (aneuploidy) lead to early embryonic death. Exceptions are aneuploidy of the sex chromosomes (e.g. Klinefelter and Turner syndromes), Down syndrome (trisomy of chromosome 21), Edwards syndrome (trisomy of chromosome 18), and Patau syndrome (trisomy of chromosome 13).

Translocations occur as a result of recombination between two non-homologous chromosomes, or within a single chromosome (for example, at sites of repeat sequences). Translocations can alter the information encoded in the genome in several ways:

- The break point may occur within a gene in a manner that destroys its function (e.g. damaged basal promoter, or truncated transcript).
- Translocated genes may come under the influence of new promoters or enhancers and have their expression pattern altered. For example, in Burkitt's lymphoma the *c-MYC* gene (normally on chromosome 8) is fused to the promoter region of the immunoglobulin genes (normally on chromosome 14), altering its pattern of expression.[19]
- The breakpoint may fuse the coding regions of two genes together, as in the Philadelphia chromosome of patients with chronic myelogenous leukemia (CML) which involves the fusion of two genes[19] (*BCR* and *ABL*).

---

[15] S Li and X-J Li, Multiple pathways contribute to the pathogenesis of Huntington disease, *Molecular Neurodegeneration*, 2006, **1**: 19.

[16] Z Jian *et al.*, Ancestral reconstruction of segmental duplications reveals punctuated cores of human genome evolution, *Nature Genetics*, 2007, **39**(11): 1361–1368.

[17] DF Conrad and ME Hurles, The population genetics of structural variation, *Nature Genetics Supplement*, 2007, **39**: S30–S36.

[18] P Stankiewicz and JR Lupski, Genome architecture, rearrangements and genomic disorders, *Trends in Genetics*, 2002, **18**(2): 74–82.

[19] Reviewed in JD Rowley, Chromosome translocations: dangerous liaisons revisited, *Nature Reviews Cancer*, 2001, **1**: 245–250.

Frequencies of DNA Sequence Variants

In humans, new point mutations are estimated to occur at an average rate of about one in every 50 million nucleotides (2×10^{-8} mutations per nucleotide per cell division). For the $\sim 6 \times 10^9$ base pairs in a diploid human cell, that means each fertilized egg will — on average — contain some 120 new point mutations.[20]

The proportion of DNA sequence affected by new structural variations (1 kb – 10 Mb) is thought to be two to four orders of magnitude higher depending on the chromosomal region.[20] For families with a history of genetic disease, the rate can be another order of magnitude higher (ibid.). Note, however, that DNA regions well inside some large structural variants (e.g. inversions) may not be functionally affected if their neighborhood is unchanged. Thus, large-scale structural variations may not always have dramatic functional consequences.

In terms of the total variability within the human genome, as of February 2009, the Database of Genomic Variants (http://projects.tcag.ca/variation/) lists 6,225 variable DNA loci, while dbSNP (http://www.ncbi.nlm.nih.gov/projects/SNP/) reports over 6.5 million single nucleotide polymorphisms. Although the number of known SNPs is much higher than the number of known structural variants, the proportion of the genome covered by structural variants is much larger. As illustrated in figure 3.4, February 2009 data from the Database of Genomic Variants estimate that on average as much as 28% of the genome is affected by structural variations (dashed horizontal line).

A previous estimate[21] put the figure at up to 600 Mb (i.e. ~20% of the genome). However, recent detailed analysis of eight individual genomes suggests both of these figures are over-estimated, perhaps by an order of magnitude, due to technical difficulties in estimating the length of large sequence variants.[22]

The above discussion centered on total genomic variability. How much difference can we expect to see between any given pair of individuals? Comparison of the individually sequenced, high-quality genome sequence of Craig Venter to the reference sequence produced by the International Human Genome Project identified about 3.2 million SNPs and an additional 9.1 Mbps of multi-nucleotide DNA variants.[23] This amounts to ~0.2% or about

---

[20] JR Lupski, Genomic rearrangements and sporadic disease, *Nature Genetics Supplement*, 2007, **39**: S43–S47.

[21] GM Cooper, DA Nickerson and EE Eichler, Mutational and selective effects on copy-number variants in the human genome, *Nature Genetics Supplement*, 2007, **39**: S22–S29.

[22] JM Kidd *et al.*, Mapping and sequencing of structural variation from eight human genomes, *Nature*, 2008, **453**: 56–64.

[23] S Levy *et al.*, The diploid genome sequence of an individual human, *PLoS Biology*, 2007, **5**(10): e254.

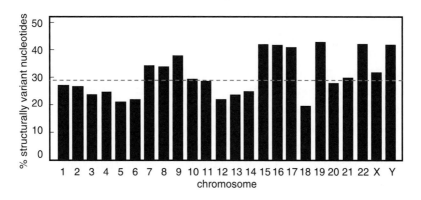

Figure 3.4: Structural variation per chromosome as reported in the Database of Genomic Variants.

one variant nucleotide in every 500 bp and is consistent with earlier estimates suggesting between 1-in-135 bp to 1-in-800 bp variation between any two individuals.[24] Analysis of James Watson's individual genome has revealed similar numbers.[25] Finally, we note that 44% of Venter's genes are affected by at least one variant, hinting at the scale and complexity of interpreting personal genomes.

Effects of DNA Sequence Variation on Cellular Biochemistry

Genetic traits can be due to single or multiple genes. In diploid organisms (such as humans), we need to consider the effect of variations within each copy of a gene.

   Dominant and recessive alleles. Some alleles have additive effects. For example, red and colorless (white) alleles in a flower may combine to produce pink petals. The red allele is then said to have incomplete dominance.[26] Many alleles are fully dominant. For example, consider two alternative alleles *G* and *g* of a transcription factor gene. Suppose *G* produces a protein that is a transcriptional silencer, while *g* produces a dysfunctional protein missing its DNA-binding domain. In a heterozygote individual bearing (*G,g*) alleles, the presence of a single

---

[24] J Sebat, Major changes in our DNA lead to major changes in our thinking, *Nature Genetics Supplement*, 2007, **39**: S3–S5.

[25] DA Wheeler, The complete genome of an individual by massively parallel DNA sequencing, *Nature*, 2008, **452**: 872–877.

[26] As we will see later, environmental and other factors can also cause incomplete penetrance.

copy of *G* will confer transcriptional silencing irrespective of the type of the second (recessive) allele. Mutations that give proteins new functions are often dominant.

In 1981, Henrik Kacser and James Burns showed that a minor allele that reduces the activity or concentration of an enzyme will have little effect on pathway flux in heterozygote individuals. Critically, Kacser and Burns showed that this is an intrinsic property of multistage metabolic and signal transduction pathways. It does not depend on the enzyme. Thus many enzyme mutations in metabolic and signaling pathways are likely to be recessive.[27]

Mendelian inheritance. In 1866 — decades before the concepts of genes, DNA, and chromosomes were accepted — Gregor Mendel showed that some traits are inherited through discrete heritable quantities (genes) which can take different values (alleles) and are independently and randomly assigned to each offspring (chromosomal assortment and crossover). Mendel's "principles of inheritance" have been the cornerstones of genetics since their re-discovery in 1900.[28]

If a trait is influenced by alleles of multiple genes, its inheritance patterns can be predicted from Mendelian principles so long as (i) each gene impacts the trait independently of the allelic values of the other genes, (ii) the alleles of different genes are assorted among the offspring independently, and (iii) all alleles are either fully dominant or fully recessive.

In figure 3.5, the top pedigree diagram shows the expected Mendelian pattern of inheritance for a dominant allele (A) from a heterozygote parent. The Punnett square to the right of the pedigree diagram shows all the combinations of the maternal and paternal alleles (gray backgrounds indicate the trait is inherited). On average, 50% of the offspring will inherit the trait. The symbols are as described in the key. F1 and F2 refer to the (Filial) generation number.

The lower panel in figure 3.5 shows an example of Mendelian inheritance of a recessive trait (e.g. cystic fibrosis, as discussed earlier in this chapter). In this case, the mother is homozygous (aa, affected), while the father is a heterozygous (Aa, unaffected) carrier.

Mendelian inheritance concepts have been of enormous value over the past century, not least by providing a theoretical framework for association studies and linkage analysis to identify disease genes. Unfortunately, most genetic diseases do not conform to Mendel's assumptions. Four key mechanisms that can lead to non-Mendelian inheritance are: chromosomal abnormalities; single-gene alleles that are neither fully dominant nor fully recessive; mutations within the X and Y chromosomes and imprinted genes; and non-additive

---

[27] H Kacser and JA Burns, The molecular basis of dominance, *Genetics*, 1981, **97**: 639–666.
[28] Mendel's original papers, and translations, are available from http://www.mendelweb.org/MWtoc.html.

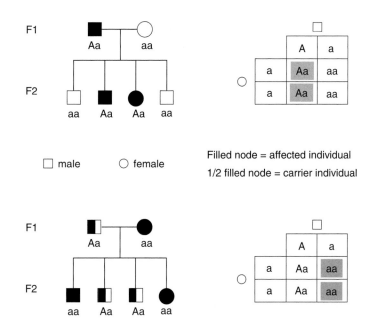

Figure 3.5: Pedigree trees and Punnett diagrams.

interactions among multiple alleles in multi-factorial diseases. Moreover, a number of processes complicate Mendelian inheritance patterns:[29]

- Co-dominance: when both alleles of a gene are expressed, as in MHC molecules.
- Incomplete Penetrance: when only a percentage of carriers develop the associated trait, as in certain types of retinoblastoma.[30]
- Variable Expressivity: when only some of all the symptoms associated with a disease-allele are expressed in any one individual. For example, Marfan syndrome (which affects collagen formation) can cause optical, skeletal and cardiovascular abnormalities, but affected individuals may present only a subset of these symptoms.
- Late/delayed onset: affected individuals may not develop the associated disease for many years, as in Huntington's disease.

---

[29] Reviewed in V van Heyningen and PL Yeyati, Mechanisms of non-Mendelian inheritance in genetic disease, *Human Molecular Genetics*, 2004, **13**(Review Issue 2): R225–R233.

[30] GA Otterson *et al.*, Incomplete penetrance of familial retinoblastoma linked to germ-line mutations that result in partial loss of RB function, *Proceedings of the National Academy of Sciences USA*, 1997, **94**: 12036–12040.

- New mutations: some diseases arise due to high rates of *de novo* mutations (e.g. Achondroplasia, a common cause of dwarfism).

- Pharmacogenetic effects: when the effect of an allele only becomes apparent in response to exogenous chemicals such as barbiturates or pharmacological drugs. For example, some men with G6PD deficiency have a hemolytic reaction (loss of red blood cells) to an ingredient in fava (broad) beans.[31]

- Incomplete Dominance: where heterozygote individuals may not exhibit symptoms except under certain conditions. For example, people heterozygote for the HbS allele in sickle cell anemia will undergo sickling when subjected to low oxygen levels.[32]

- Mutations affecting mitochondrial function: health outcomes of mutations in mitochondrial DNA depend on the number of mitochondria affected. In mice, an embryo inherits about 200,000 mitochondria in the maternal egg. This number of mitochondria is thought to be amplified from a subset of between 20 to 2,000 maternal mitochondria in the egg progenitor cells.[33] As a result, mothers transmit a varying proportion of their mitochondrial mutations to each offspring. Thus different offspring may inherit different types of alleles (as well different loadings). Mitochondrial mutations affect many tissues and can manifest as heart failure, stroke-like episodes, and several neuromuscular diseases.[34]

Multi-gene traits. Most sequence variants have a very small effect on the associated trait, indicating that additional genes and/or environmental factors also influence the outcome. For example, variants in the *PPARG* and *CAPN10* genes are predictive of future development of type 2 diabetes mellitus.[35] However, the combined predictive power of these variants is not significantly higher than predictions based on Body Mass Index (BMI) and fasting plasma glucose (FPG) levels.[36] In a similar vein, a common variant of the (tumor-associated)

---

[31] G6PD has many alleles, only some of which cause Favism; see http://www.g6pd.org/favism.

[32] RM Bookchin, T Balazs and LC Landau, Determinants of red cell sickling. Effects of varying pH and of increasing intracellular hemoglobin concentration by osmotic shrinkage, *Journal of Laboratory and Clinical Medicine*, 1976, **87**(4): 597–616.

[33] LM Cree *et al.*, A reduction of mitochondrial DNA molecules during embryogenesis explains the rapid segregation of genotypes, *Nature Genetics*, 2008, **40**: 249–254. K Khrapko, Two ways to make an mtDNA bottleneck, *Nature Genetics*, 2008, **40**: 134–135.

[34] See for example A Chomyn, The myoclonic epilepsy and ragged-red fiber mutation provides new insights into human mitochondrial function and genetics, *American Journal of Human Genetics*, 1998, **62**: 745–751.

[35] V Lyssenko *et al.*, Genetic prediction of future type 2 Diabetes, *PLoS Medicine*, 2005, **2**(12): e345.

[36] ACJW Janssens *et al.*, Does genetic testing really improve the prediction of future type 2 diabetes? *PLoS Medicine*, 2006, **3**(2): e114.

HMGA2 gene has a significant statistical association with both childhood and adult height. However, this variant only explains about 0.3% of the population variation in height and accounts for only about 0.4 cm increased adult height per allele,[37] suggesting that multiple additional factors and nonlinear interactions may be involved.

There are two distinct ways in which multiple genetic loci may contribute to a given disease. One possibility is that each genetic variant can individually and independently cause the disease. In that case, each allele would follow Mendelian inheritance patterns. According to this model, common genetic diseases arise from rare occurrences of a wide variety of mutations. This is thought to be the case for some cases of schizophrenia,[38] while in other cases, schizophrenia may be due to co-occurrences of common variants.[39]

To explore the concept of combinatorial genetic effects, consider the minimal case with just two genes. Suppose gene 1 has alleles *G1* and *g1*; and gene 2 has alleles *G2* and *g2*. *G1* and *G2* are disease-associated, dominant over *g1* and *g2*, and act independently. Figure 3.6 summarizes the disease traits expected. The four corners of the square represent the four possible allelic combinations in a diploid individual (indicated in brackets inside the square). Colored background boxes mark the disease-associated combinations.

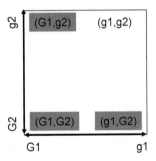

Figure 3.6:   Phenotypes of two dominant alleles.

---

[37] MN Weedon *et al.*, A common variant of *HMGA2* is associated with adult and childhood height in the general population, *Nature Genetics*, 2007, **39**(11): 1245–1250.

[38] T Walsh *et al.*, Rare structural variants disrupt multiple genes in neurodevelopmental pathways in schizophrenia, *Science*, 2008, **320**(5875): 539–543; H Stefensson *et al.*, Large recurrent microdeletions associated with schizophrenia, *Nature*, 2008, **455**(7210): 178–179; The International Schizophrenia Consortium, Rare chromosomal deletions and duplications increase risk of schizophrenia, *Nature*, 2008, **455**(7210): 237–241.

[39] The International Schizophrenia Consortium, Common polygenic variation contributes to risk of schizophrenia and bipolar disorder, *Nature*, 2009, **460**(7256): 748–752.

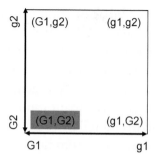

Figure 3.7: Incomplete penetrance of two alleles.

The second way in which two or more loci may contribute to a trait is that only particular combinations of alleles at different sites cause the trait of interest. In that case, *G1* and *G2* may be said to have incomplete penetrance. Figure 3.7 gives an example combinatorial effect. Here, a disease only arises for the allele combination (*G1,G2*). This could be the case if, for example, *G1* and *G2* are dysfunctional, whereas *g1* and *g2* can each independently activate transcription of an important downstream target gene (e.g. a tumor suppressor).

In combinatorial multi-gene traits, individual alleles each make a modest statistical contribution to the disease phenotype. Given that biochemical processes in human cells (e.g. transcriptional regulation) are typically robustly regulated by many gene products, this model would predict that many disease phenotypes will not arise except in the presence of *particular combinations* of multiple alleles.[40] In this way, relatively common alleles could give rise to much less frequent diseases (an example of the "common-disease, common-allele" paradigm). Note, however, that in this scenario the individual alleles have very small effect sizes and may not be detectable independently; they will be detectable only when considered in combination.

Instead of sharply defined combinatorial effects, interacting alleles of multiple genes may simply contribute additively to a phenotype, as illustrated in figure 3.8. In this case, heterozygotes will exhibit a milder form of the homozygous trait, and the trait is said to be quantitative. In the figure, lighter shade background boxes indicate a partial phenotype, which may be different for each heterozygote combination. An example would be where the

---

[40] The concept of combinatorial multi-gene traits is discussed in the context of plant genetics in DJ Kliebenstein, Advancing genetic theory and application by metabolic quantitative trait loci analysis, *Plant Cell*, 2009, **21**:1637–1646.

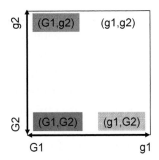

Figure 3.8: Quantitative combinatorial effects.

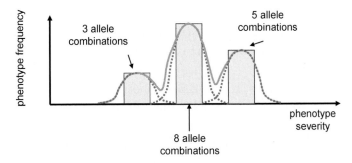

Figure 3.9: Example of a complex relationship between allele frequency and phenotype distribution.

minor alleles of one gene predispose carriers to an illness while alleles of another gene protect against that disease to varying degrees.[41]

Combinatorial effects can complicate the relationship between allele frequency and phenotype frequency, as illustrated in figure 3.9. In this example, a trait is assumed to be the outcome of interactions among four bi-allelic genes. Thus, there are 16 possible genotypes (2^4). Suppose that all the individual alleles occur with equal frequency, and that the 16 genotypes cluster into three distinct phenotypes, as indicated by the solid bars and labels in the figure. We see that in this case, the frequency of each observed phenotype is proportional to the number of allele combinations that generate the phenotype, not the allele frequency.

Variations in environmental and other influences can make discrete phenotype distributions (as in the solid bars above) appear continuous. In the above figure the three dashed

---

[41] JH Nadeau and EJ Topol, The genetics of health, *Nature Genetics*, 2006, **38**(10): 1096–1098.

bell-shaped curves illustrate the effect of non-genetic factors on the three genetic pheno-types. The triple-peaked solid curve indicates the total observed phenotype frequency, which appears continuous. Larger numbers of distinct phenotypes arising from different allelic com-binations can make the phenotypic distribution appear more continuous.

Consider the case when two genes acting on a trait each have multiple functionally dis-tinct alleles (for example, sequence variations affecting the stability, alternative splicing, and interaction domains of the gene products). In that case, both additive and combinatorial effects can lead to a distribution of trait values. In theory this distribution can be discrete, but in practice random variability is likely to blur the discrete phenotypes into a range, as illustrated in figure 3.10. Here, genes *G1* and *G2* are each assumed to have N alleles of varying severity. Genotypes (allele combinations) are organized such that the more severe phenotypes (darker shading) occur nearer the bottom-left corner of the square. In this illustrative plot the phenotype severity is symmetrically distributed around the 45 degree diagonal. In general, this would not be the case and arbitrary distributions of severity may arise.

The three-dimensional images in figure 3.11 show example trait-intensity distributions for alleles that interact additively (top-left panel) or combinatorially (remaining panels). Here, the phenotype intensity (health/disease status) is indicated along the vertical axis. In the top-left panel, the shaded surface illustrates an additive effect of the $(G1, G2)$ alleles. In this example, the two alleles make equal contributions to the phenotype.

The phenotype in the top-right panel also has equal contributions from the two alleles, but the step-like shaded surface illustrates combinatorial action by the $(G1, G2)$ alleles, i.e. only specific combinations of the two alleles cause a disease phenotype. The two lower panels show examples of unequal combinatorial action by two alleles. In both plots, the alleles of *G2* have been arranged such that more healthy phenotypes are positioned to the

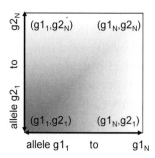

Figure 3.10: Interactions of two genes each with many alleles.

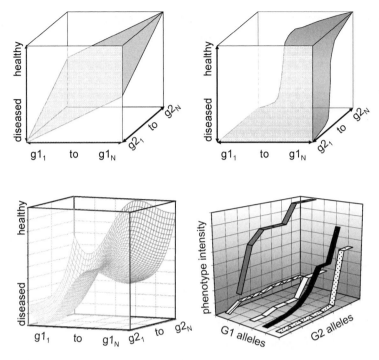

Figure 3.11:   Four types of interaction between alleles of two genes.

right and back of the page. We see that for a given *G2* allele, different *G1* alleles can lead to varying degrees of health/disease.

In the bottom-left panel, *G1* alleles interact similarly with all *G2* alleles (*G1* is said to modulate *G2*). In the bottom-right panel, each *G1* allele interacts with the *G2* alleles in a different manner (indicated by the distinct shapes of the curves).

Combinatorial effects involving multiple alleles of multiple genes complicate statistical searches for allele-disease correlations. The same allele will be associated with a disease in some individuals but not others, depending on allelic differences elsewhere in the genomes. Although this fact has been known from the earliest days of genetics,[42] it has only become possible to address it in recent years — as we will see in the rest of this book.

How prevalent or rare are multi-allele and combinatorial effects? One way to measure the effects of allelic variations is to measure the extent to which gene expression patterns vary

---

[42] PC Phillips, The language of gene interaction, *Genetics*, 1998, **149**: 1167–1171.

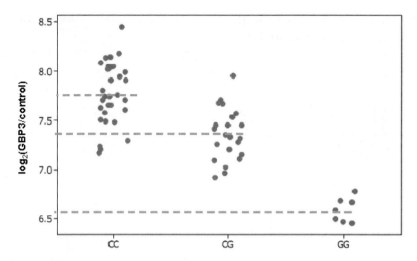

Figure 3.12:   Relationship between allelic variation and gene expression levels.

among individuals within a genetic population. For example, figure 3.12 shows statistically significant variation in *GBP3* gene expression associated with three haplotype variants among 60 Utah residents of European descent.[43] Each disk represents the mRNA transcript levels (relative to a control) in an individual (averaged over four technical replicates). Each column of scattered plot points represents one allelic variant. The dashed green lines indicate the average mRNA abundance per allele.

There is a 1.5 to 2.5-fold variation in expression between individuals in different allelic groups. Remarkably, the variation within allelic groups is as large as the inter-allele variation, suggesting multiple additional factors influence mRNA abundance differences between individuals.[44]

The above conclusions are supported by a study of the extent to which genetic background affects gene expression levels in five common laboratory strains of mice.[45] About 13%

---

[43] Reproduced from figure 2 of BE Stranger *et al.*, Relative impact of nucleotide and copy number variation on gene expression phenotypes, *Science*, 2007, **315**: 848–853. Reprinted with permission from AAAS.

[44] See also JD Storey *et al.*, Gene-expression variation within and among human populations, *The American Journal of Human Genetics*, 2007, **80**: 502–509.

[45] D Bianchi-Frias *et al.*, Genetic background influences murine prostate gene expression: implications for cancer phenotypes, *Genome Biology*, 2007, **8**: R117.

of genes expressed in the prostate (932 genes) exhibited differential expression in the range 1.3 to 190-fold between any two strains.

It has been estimated that for any two individuals, between 17% and 29% (roughly one sixth to one third) of all genes will be expressed at significantly different levels.[46] In another study, expression levels of 28% of 20,599 genes measured in immortalized B cells of 400 children from 206 families were found to be significantly correlated with parental gene expression levels.[47] For all three of the preceding studies, the expression variability was highly correlated with DNA sequence variation within 100 kb of the affected gene.

The above observations about mRNA abundance variability are supported by studies that also measured protein levels. In yeast, a controlled experiment in which two different strains of yeast were mated and parent–offspring protein abundances compared, found that more than one third of all proteins differed significantly in abundance between allelic groups.[48] The average difference between segregants was about two-fold. In humans, a study of 50 individuals from different ethnic and geographical backgrounds found that increased numbers of copies (up to 16) of the amylase gene (needed for starch digestion) are selected in populations with starch-rich diets. The increased copy numbers were shown to correlate directly with amylase protein concentration in the saliva.[49]

Whether genetic variations in mRNA and protein concentrations affect cellular function will depend on the interaction networks that the affected biomolecules take part in. As we will see in Chapter 5, network behavior can be affected in diverse ways depending on which network component is mutated and functional effect of the mutation on the affected molecular species. Thus, whether the presence of one or more alleles will cause a trait will be tissue and condition dependent.

In short, except for relatively rare, independently acting Mendelian alleles, the effect of any given allele on cellular physiology will depend on the particular combination of other alleles present in the network of molecular interactions in which that allele products take part. These considerations suggest that for many diseases it will be necessary to use pathway information and dynamic behavior models to predict the effects of multiple, interacting genomic

---

[46] BE Stranger *et al.*, Population genomics of human gene expression, *Nature Genetics*, 2007, **39**(10): 1217–1224.

[47] AN Dixon, A genome-wide association study of global gene expression, *Nature Genetics*, 2007, **39**(11): 1202–1207.

[48] EJ Foss *et al.*, Genetic basis of proteome variation in yeast, *Nature Genetics*, 2007, **39**(11): 1369–1375.

[49] GH Perry *et al.*, Diet and the evolution of the human amylase genes copy number variation, *Nature Genetics*, 2007, **39**(11): 1256–1260.

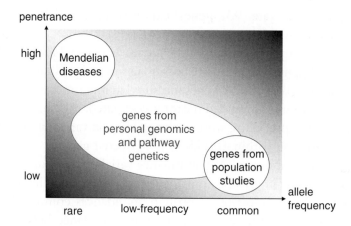

Figure 3.13:   The relationship between penetrance and allele frequency.

sequence variations. In contrast to the Mendelian assumption of independently acting alleles, each individual may have a unique combination of predisposing and protective alleles interacting in complex, nonlinear ways. However, all affected individuals may share common dysregulated molecular pathways and processes. Figure 3.13 indicates the relationship between this new, system-oriented approach to genetics and traditional perspectives.

In the above figure, the vertical axis (penetrance) represents the proportion of allele carriers who are symptomatic. The darker areas towards the top-right and bottom-left corners represent unlikely scenarios for genetic diseases. Common alleles that cause widespread disease will be selected against. Rare alleles that are weakly associated with symptoms suggest the disorder is primarily caused by environmental factors.

Alleles of single-gene and multi-gene Mendelian diseases (in which alleles contribute to the phenotype independently) exhibit high penetrance. These alleles are typically selected against over evolutionary timescales, so they tend to be rare in populations (indicated at the top left of the figure). Nonetheless, studies using family trees and other well-established approaches have identified the tractable cases. Those that remain will require whole-genome sequencing of affected individuals to identify mutations in unexpected DNA regions.

At the other extreme (indicated at the bottom-right of the plot) are commonly occurring alleles that are individually very poor predictors of whether a carrier will become symptomatic.

Large-scale population studies have identified many such variations.[50] However, for this class of genetic diseases, interactions with other alleles and environmental factors clearly play an important role in determining the phenotypic outcome. So understanding the molecular pathways within which these alleles operate will be crucial to predicting their effect.

Between the above two categories lies a broad spectrum of alleles and diseases that are difficult to study through traditional approaches. Typically these are diseases that arise from the combinatorial interactions of multiple, low-frequency and rare alleles with each other and with environmental factors.

Finding disease-associated combinations of alleles in population data is statistically challenging. Every combination considered amounts to a new hypothesis being tested. Because of the need to correct for multiple-hypothesis testing, current population-based approaches can only detect cases arising from interactions of at most two alleles, not three or more combinations.[51]

To illustrate the challenge, consider genome-wide SNP association studies. Scanning all genomic SNPs for possible statistical associations to diseases is equivalent to testing for about 1 million hypotheses.[52] To detect a significant correlation in the presence of such multiple-hypothesis testing, p-values below 10^{-8} are needed.[53] Large population sizes are required to avoid apparent associations that are later disproved.

Figure 3.14 highlights the need to go beyond traditional genetics in order to discover multi-factorial low-frequency disease alleles.[53] Here, the horizontal axis represents the fold increase in disease odds for carriers compared to non-carriers. The vertical axis indicates the number of people required for a "classical" case-control genetic association study to find

---

[50] MI McCarthy *et al.*, Genome-wide association studies for complex traits: consensus, uncertainty and challenges, *Nature Reviews Genetics*, 2008, **9**: 356–369.

[51] J Marchini, P Donnelly and LR Cardon, Genome-wide strategies for detecting multiple loci that influence complex diseases, *Nature Genetics*, 2005, **37**(4): 413–417; DM Evans *et al.*, Two-stage two-locus models in genome-wide association, *PLoS Genetics*, 2006, **2**(9): e157; Y Zhang and JS Liu, Bayesian inference of epistatic interactions in case-control studies, *Nature Genetics*, 2007, **39**(9): 1167–1173.

[52] I Pe'er *et al.*, Estimation of the multiple testing burden for genomewide association studies of nearly all common variants, *Genetic Epidemiology*, 2008, **32**(4): 381–385.

[53] Figure reproduced from D Altshuler, MJ Daly and ES Lander, Genetic mapping in human disease, *Science*, 2008, **322**: 881–888. Reprinted with permission from AAAS.

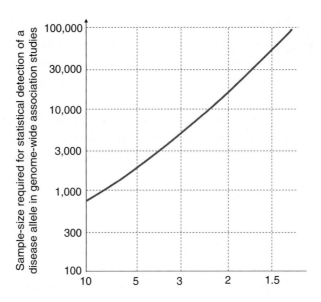

Figure 3.14: Required sample size as a function of fold change in disease odds given a marker.

markers in the genome correlated with the disorder and assuming no prior information (such as candidate genes or pathways).

The plotted curve is for a disease allele that occurs, on average, in one out of every 100 individuals. Rarer alleles will require larger samples. As expected, the required sample size (the number of cases plus the number of controls) increases sharply for alleles with less pronounced effects on disease odds.

Figure 3.15[54] shows the distribution of fold increase in disease odds (odds ratios) for 61 rare variants (minor allele frequency less than a few percent) and 217 common variants reported in the literature. We see that almost all common variants (dark bars), and many rare variants (white bars) have odds ratios less than 2.

Referring back to figure 3.14, we note that population studies of variants with odds ratios ≤2 will require populations of 20,000 or more cases, at enormous cost. In such cases, studies using candidate genes and selected case groups would be far more efficient at finding the genetic culprits (discussed in Ref. 54). Traditionally this has not been possible because of the

---

[54] Reprinted with permission from Macmillan Publishers Ltd: W Bodmer and C Bonilla, Common and rare variants in multifactorial susceptibility to common diseases, *Nature Genetics*, 2008, **40**(6): 295–701. Copyright (2008).

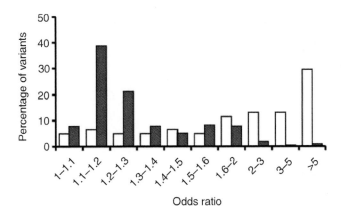

Figure 3.15: Distribution of disease odds ratios for some rare and common variants.

need for extensive DNA re-sequencing of many genes in large numbers of individuals, and the assessment of the functional consequences of the identified variants. However, as we will see in the rest of this book, the emergence of low-cost, high-throughput DNA sequencing technologies, and systems biology approaches to candidate gene selection are removing these barriers, and opening the doors to a new era in genetics.

Personal genomics will alleviate the increasing cost and complexity of statistical association studies. Full genomic sequencing and detailed molecular characterization of dysregulated pathways and processes in multiple affected individuals will indicate commonly affected pathways and processes. The components of common dysregulated pathways provide excellent candidate genes for rarer alleles. We will return to the topic of pathway-oriented, systems genetics repeatedly throughout this book.

CHAPTER 4

Environmental and Life-History Effects

The previous chapter focused on variability in DNA sequence among individuals, and its effects on cellular and organ physiology. This chapter reviews how biochemical and physiological differences among individuals also arise from differences in environmental exposures and life history.

Most adverse health events are caused by a combination of genetic and environmental factors, but the extent to which genetic and environmental factors are causal varies. Setting risk-taking behaviors aside, random accidents do not usually have a genetic cause, but people may have differing genetic predispositions to injury (e.g. brittle bones). For contagious diseases, environmental and behavioral factors are causal. But people from different genetic backgrounds can have different susceptibilities to contagious diseases.[1] In contrast to the preceding examples, some "simple genetic diseases" such as Huntington's are highly predictable from genetic data alone.

The relationship between genes, environment, and health is visualized schematically in figure 4.1.[2] At the top right are the diseases that pose the greatest challenges in the technologically developed world. These "complex diseases" are multi-factorial, i.e. they arise from interactions among multiple genetic and environmental factors. Examples of such disorders are cardiovascular diseases, cancers, allergies, asthma, diabetes, multiple sclerosis, Alzheimer's and so on. This chapter describes some of the mechanisms by which environmental effects can influence the regulation of gene expression, and therefore cellular function, organ physiology and health status.

A key take-home message of this chapter is that, for most people and diseases, genomic sequence alone can only act as a partial, probabilistic predictor of health status. For greater

---

[1] See for example P-Y Bochud *et al.*, Polymorphisms in Toll-Like Receptor 4 (TLR4) are associated with protection against leprosy, *European Journal of Clinical Microbiology and Infectious Diseases*, published online 09 May 2009.

[2] I thank Dr. Gustavo Glusman (Institute for Systems Biology, Seattle, USA) for suggesting this figure.

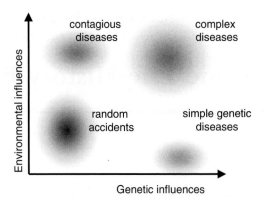

Figure 4.1:   Health impacts of environmental and genetic factors.

predictability of health status, we will need to also assay DNA and chromatin modifications in particular cells and organs, and measure metabolite, mRNA, lipid and protein levels and modification states in the affected organs (discussed further in Chapters 6 and 7).

Somatic Mutations in Cancer

We noted in Chapter 3 that many inherited DNA variations (e.g. mutations in the tumor suppressor gene P53) can predispose individuals to particular forms of cancer. But most cancers arise from an accumulation of multiple somatic-cell mutations during the lifetime of an individual. As of June 2009, the Catalogue of Somatic Mutations in Cancer (COSMIC) database[3] (http://www.sanger.ac.uk/genetics/CGP/cosmic/) lists 78,933 cancer-associated somatic mutations in 4,775 genes.

It is increasingly apparent that individuals with the same type of cancer often harbor different types of cancer-associated somatic mutations. For example, two recent studies of glioblastoma and one of pancreatic cancer uncovered distinct sets of dozens of sequence changes per patient.[4]

---

[3] C Bamford *et al.*, The COSMIC (Catalogue of Somatic Mutations in Cancer) database and website, *British Journal of Cancer*, 2004, **91**: 355–358.

[4] The Cancer Genome Atlas Research Network, Comprehensive genomic characterization defines human glioblastoma genes and core pathways, *Nature*, 2008, **455**(7216): 1061–1068; DW Parsons *et al.*, An integrated genomic analysis of human glioblastoma multiforme, *Science*, 2008, **321**(5897): 1807–1812; S Jones *et al.*, Core signaling pathways in human pancreatic cancers revealed by global genomic analyses, *Science*, 2008, **321**(5897): 1801–1806.

Interestingly, though distinct, most of the mutations affected a common set of pathways. For example, over two thirds of pancreatic cancer tumors (average of 63 somatic mutations per tumor) disrupted 12 shared "core" pathways.

Known differences in the complement of mutations harbored by individual cancer patients can be exploited to provide targeted therapies. For example, about 10% of Western European and American patients with non small cell lung cancer have a particular gain-of-function mutation in their epidermal growth factor receptor (EGFR) gene (the proportion is higher in east Asians). EGF-pathway activity-suppressor drugs such as gefitinib can be an effective treatment in these patients.[5]

Lung cancer is a good example of how environmental influences affect cancer risk. About 85% of lung cancers are caused by smoking.[6] Smoking has been shown to result in widespread gene expression changes, some of which are not reversed in ex-smokers.[6] Among the affected genes is GSK3β (Glycogen Synthase Kinase 3 beta), whose expression is irreversibly reduced in smokers and ex-smokers. GSK3β is an enzyme that interacts with more than 30 different substrates in different cells and conditions. Thus, the effects of lowered GSK3β levels may be widespread. Importantly, GSK3β activity is inhibited during canonical Wnt signaling, and inappropriate canonical Wnt signaling is implicated in cancer stem cells.[7] Moreover, GSK3β inhibits the transcription of *Cox2* (Cyclooxygenase-2), an inflammatory-response gene[8] involved in a positive feedback loop with EGFR and implicated in many tumors.[9] Thus, lowered GSK3β levels caused by smoking may predispose former and current smokers to multiple cancer types.

Among uranium ore miners[10] and nuclear plant workers,[11] radiation is a major cause of lung cancer, presumably through inhalation of trace amounts of radioactive material.

---

[5] SV Sharma *et al.*, Epidermal growth factor receptor mutations in lung cancer, *Nature Reviews Cancer*, 2007, 7: 169–181.

[6] R Chari *et al.*, Effect of active smoking on the human bronchial epithelium transcriptome, *BMC Genomics*, 2007, **8**: article 297.

[7] T Reya and H Clevers, Wnt signalling in stem cells and cancer, *Nature*, 2005, **434**: 843–850.

[8] A Thiel *et al.*, Expression of cyclooxygenase-2 is regulated by Glycogen Synthase Kinase-3β in gastric cancer cells, *Journal of Biological Chemistry*, 2006, **281**(8): 4564–4569.

[9] MS Choe *et al.*, Interaction between epidermal growth factor receptor- and cyclooxygenase 2-mediated pathways and its implications for the chemoprevention of head and neck cancer, *Molecular Cancer Therapeutics*, 2005, **4**(9): 1448–1455.

[10] JW McDonald *et al.*, p53 and K-ras in radon-associated lung adenocarcinoma, *Cancer Epidemiology, Biomarkers and Prevention*, 1995, **4**(7): 791–793.

[11] CM Lyon *et al.*, Radiation-induced lung adenocarcinoma is associated with increased frequency of genes inactivated by promoter hypermethylation, *Radiation Research*, 2007, **168**: 409–414.

Example mechanisms of radiation-induced lung cancer include DNA sequence point mutations[10] and epigenetic modifications of DNA[11] (promoter sequence methylation). Thus, two very different environmental factors (smoking and radiation) can result in the same disease (lung cancer) via different mechanisms.

Cancer offers a particularly compelling and well-studied example of how environmental effects (exposure to carcinogens) and life history (including aging) lead to dramatic biochemical differences between individuals, and how personalized treatments can address specific molecular disorders in particular patients. But cancer encompasses a broad range of diseases and involves a multi-step, highly complex sequence of dysregulation events. Thus, the interactions of "nature and nurture" in cancer may appear to some readers to be a special case. To avoid this impression, the remainder of this chapter will pay greater attention to environmental and life-history effects relating to other diseases. For readers especially interested in cancer, many books and articles describe and discuss specific aspects of cancer.[12]

Epimutations

We noted in Chapter 2 that regulation of gene expression can occur through "epigenetic" chemical modifications of histones and DNA (e.g. methylation). A remarkably wide variety of environmental conditions can give rise to epigenetic modifications. Here we review just a few examples to highlight environmental modulation of genetic regulatory programs.

Many environmental factors have been found to change DNA methylation patterns. CG dinucleotides are relatively uncommon in the human genome. But in the sequences flanking transcription start sites, CG dinucleotides occur much more frequently, forming relatively CG-rich regions of a few hundred to a few thousand nucleotides known as CpG islands (the "p" denotes the phosphodiester bond between the C and G nucleosides). As noted in Chapter 2, cytosine methylation tends to repress transcription.

Most CpG islands are not methylated in humans, but they can become methylated as a result of specific experiences. For example, it has been shown that rat pups receiving greater grooming and care from their mothers in the first week after birth, have de-methylated CG dinucleotides in a hippocampal glucocorticoid receptor (GR) gene promoter.[13] In uncared-for

---

[12] As of August 2008, Amazon.com listed 590 books tagged as "cancer" related. For a recent review article, see L Chin and JW Gray, Translating insights from the cancer genome into clinical practice, *Nature*, 2008, **452**: 553–563. For a general review of cancer biology, see The Biology of Cancer by Robert A. Weinberg, Garland Science, 2006.

[13] ICG Weaver *et al.*, Epigenetic programming by maternal behavior, *Nature Neuroscience*, 2004, 7(8): 847–854.

pups, this region is methylated and the GR gene is suppressed, resulting in more pronounced hypothalamic "fearful" responses to stress signals.

As discussed in Chapter 3, nearly half the human genome comprises transposable elements (TEs). Many TEs have important roles in facilitating genomic organization[14] and evolution.[15] Most TEs in the human genome are inactive, but a variety of environmental signals can activate some TEs.[16]

About a quarter of the proximal promoter regions in the human genome contain TEs.[17] Many TEs include a promoter capable of driving the transcription of nearby genes. In addition, most TEs are flanked by repetitive sequences that can give rise to small interfering RNAs that disrupt the expression of target genes and other TEs.[18]

The best studied and most famous example of TE-mediated epigenetic regulation is mosaic change of coat color (yellow and brown) in mice with a particular allele of the *agouti* gene (called viable yellow agouti or A^{vy}). The A^{vy} allele is the result of the insertion of a transposable element about 100 kb upstream of the transcriptional start site of the *agouti* gene. A promoter in the proximal end of this TE promotes constitutive ectopic *agouti* transcription, leading to yellow fur, diabetes, obesity and tumorigenesis. It has been shown that feeding female mice with a diet rich in methyl groups prior to and during pregnancy changes CpG sites in the *agouti*-associated TE region in the offspring from unmethylated (yellow coat) to methylated (brown coat) in a graded manner.[19] Because these epigenetic changes occur in both somatic and the germ cells of the offspring, they are also present in the second-generation offspring.[20]

---

[14] PE Warburton *et al.*, Nonrandom localization of recombination events in human alpha satellite repeat unit variants: implications for higher-order structural characteristics within centromeric heterochromatin, *Molecular and Cellular Biology*, 1993, **13**(10): 6520–6529.

[15] MT Romanish *et al.*, Repeated recruitment of LTR retrotransposons as promoters by the anti-apoptotic locus NAIP during mammalian evolution, *PLoS Genetics*, 2007, **3**(1): e10.

[16] Reviewed in RL Jirtle and MK Skinner, Environmental epigenomics and disease susceptibility, *Nature Reviews Genetics*, 2007, **8**: 253–262.

[17] IK Jordan *et al.*, Origin of a substantial fraction of human regulatory sequences from transposable elements, *Trends in Genetics*, 2003, **19**(2): 68–72.

[18] For a general review of TEs, see RK Slotkin and R Martienssen, Transposable elements and the epigenetic regulation of the genome, *Nature Reviews Genetics*, 2007, **8**: 272–285.

[19] RA Waterland and RL Jirtle, Transposable elements: targets for early nutritional effects on epigenetic gene regulation, *Molecular and Cellular Biology*, 2003, **23**(15): 5293–5300.

[20] RA Waterland, M Travisano and KG Tahiliani, Diet-induced hypermethylation at agouti viable yellow is not inherited transgenerationally through the female *FASEB Journal*, 2007, **21**: 3380–3385; JE Cropley *et al.*, Germ-line epigenetic modification of the murine A^{vy} allele by nutritional supplementation, *Proceedings of the National Academy of Sciences*, 2006, **103**(46): 17308–17312.

The proportion of functional TEs in human genomes appears to be highly variable among individuals, even within genetic groups.[21] In addition to sequence variability, there is considerable inter-individual variability in the CG methylation patterns of the most common TE family in the human genome.[22] Thus, TE-mediated epigenetic effects may vary in nature and extent from individual to individual. For instance, colorectal cancer can arise from inherited (germline) DNA sequence mutations in the mismatch repair genes *MLH1 and MSH2*. Promoter methylation of one copy of *MLH1* (and therefore its transcriptional silencing) predisposes individuals to colorectal cancer, while methylation of both copies of *MHL1* leads to sporadic (i.e. non-inherited) colorectal cancer.[23]

To emphasize the prevalence of environmental effects on the epigenome, let us briefly consider two more examples of differential DNA methylation by environmental factors. First, ambient air pollution (in this case at steel plants near a highway) has been shown to increase long-lasting global DNA methylation rates in the sperm of male mice exposed for ten weeks.[24] Second, the oncogenic adenovirus protein e1a induces quiescent human cells to replicate. In infected cells the e1a protein selectively binds to the promoters of cell cycle and growth genes, causing local histone hyper-acetylation and transcriptional activation.[25]

Inheritance of epigenetic information has been much debated in recent years. In humans, epigenetic marks on parental genomes are erased in two waves. The first time is shortly after fertilization and through the first few cell divisions, when rapid, large-scale demethylation of the sperm genome and gradual, but equally global demethylation of the egg genome occur.

The second wave of demethylation occurs specifically in the developing germ cells. Recall that in humans (primordial) germ cells are formed in the inner cell mass of the blastocyst

---

[21] M del Carmen Seleme *et al.*, Extensive individual variation in L1 retrotransposition capability contributes to human genetic diversity, *Proceedings of the National Academy of Sciences*, 2006, **103**(17): 6611–6616.

[22] I Sandovici *et al.*, Interindividual variability and parent of origin DNA methylation differences at specifc human Alu elements, *Human Molecular Genetics*, 2005, **14**(15): 2135–2143.

[23] MP Hitchins *et al.*, Inheritance of a cancer-associated MLH1 germ-line epimutation, *New England Journal of Medicine*, 2007, **356**: 697–705; ST Yuen *et al.*, Germline, somatic and epigenetic events underlying mismatch repair defciency in colorectal and HNPCC-related cancers, *Oncogene*, 2002, **21**: 7585–7592; CM Suter, DIK Martin and RL Ward, Germline epimutation of MLH1 in individuals with multiple cancers, *Nature Genetics*, 2004, **36**(5): 497–501.

[24] C Yauk *et al.*, Germ-line mutations, DNA damage, and global hypermethylation in mice exposed to particulate air pollution in an urban/industrial location, *Proceedings of the National Academy of Sciences USA*, 2008, **105**(2): 605–610.

[25] R Ferrari *et al.*, Epigenetic reprogramming by adenovirus e1a, *Science*, 2008, **321**: 1086–1088.

during week 2 of development. Gastrulation occurs shortly after the formation of the germ cells (around day 16). Two weeks later, the germ cells begin to migrate, arriving in the newly formed gonadal ridge by the end of week 5. From here on, most cells are subjected to lineage-specific *de novo* DNA methylation. However, after arriving at the gonadal ridge, the primordial germ cells undergo a unique, second demethylation wave.[26] Nonetheless, it appears that sometimes the two waves of demethylation do not erase *all* parental DNA methylation in germ cells, resulting in potential epigenetic inheritance.

A dramatic example of inherited epigenetic effects was observed in the male offspring of gestating female rats treated with either of two hormone disruptors (a fungicide and a pesticide).[27] High-dose treatment during the period when germ cells in the embryo normally undergo gender-specific DNA methylation led to hypo and hyper DNA methylation, reduced sperm count and motility, and increased male infertility. Remarkably, the effects were passed on through the male germline over four generations.[28]

We discussed the multi-generation epigenetic effects of diet on *agouti* earlier. A further example of epigenetic inheritance involving nutrition (in this case in humans) is given at the end of this chapter.[79] About half a dozen other epigenetic heritable traits have been reported in recent years.[29] Most of these studies are in model organisms.

In humans, probably the best-documented case is the effect of the estrogen diethylstilbestrol (DES) on pregnant women. Although two studies in the 1950s suggested that DES was not effective as an anti-miscarriage treatment, it continued to be prescribed for this purpose until adverse effects emerged in 1971. Since then studies have found increased risk of breast cancer in mothers, and increased risks of cancer, fertility and other genital tract related diseases in both male and female offspring exposed *in utero*.[30] Recent partial evidence

---

[26] These steps are reviewed in C Allegrucci *et al.*, Epigenetics and the germline, *Reproduction*, 2005, **129**: 137–149. For a more up-to-date review focusing purely on mouse data, see H Sasaki and Y Matsui, Epigenetic events in mammalian germ-cell development: reprogramming and beyond, *Nature Reviews Genetics*, 2008, **9**: 129–140.

[27] MD Anway *et al.*, Epigenetic transgenerational actions of endocrine disruptors and male fertility, *Science*, 2005, **308**: 1466–1469.

[28] It is not clear how the reported epigenetic transmission could have occurred down more generations than can be explained by *in utero* perturbation of germ cells in a developing embryo.

[29] Reviewed in PD Gluckman, MA Hanson and AS Beedle, Non-genomic transgenerational inheritance of disease risk, *BioEssays*, 2007, **29**: 145–154.

[30] Reviewed in M Veurink *et al.*, The history of DES, lessons to be learned, *Pharmacy World Science*, 2005, **27**: 139–143.

suggests that these epigenetic effects may be transmitted as far as third-generation offspring via sons of DES-treated mothers.[31]

The Metagenome

In addition to differences in DNA sequence and chromatin state, individuals are fundamentally different at the cellular and biochemical levels in at least three aspects: the nervous system, the immune system, and the digestive system.

Much of the development of the human brain occurs after birth. Studies of visual[32] and auditory[33] development suggest that human developmental reorganization of sensory pathways may extend over several years. Interestingly, some of the post-natal development in the cortex appears to be epigenetically regulated.[34]

Thus, human brains are necessarily differently wired in each individual. Moreover, the complex process of maturation of the human nervous system is vulnerable to dysregulation in a wide variety of ways. For example, in the absence of compensatory training, congenital visual impairment can lead to a variety of brain developmental deficits affecting cognitive and behavioral functions.[35]

The immune systems of any two individuals differ in multiple ways.[36] First, the Human Major Histocompatibility Complex or MHC locus on chromosome 6 carries some of the most polymorphic genes in the human genome and has been implicated in many autoimmune and inflammatory diseases.[37] MHC class I molecules are on the surface of all nucleated cells in the body, while class II MHC molecules are specific to immune cells.

---

[31] MM Brouwers *et al.*, Hypospadias: a transgenerational effect of diethylstilbestrol? *Human Reproduction*, 2006, **21**(3): 666–669.

[32] BM Hooks and C Chen, Critical periods in the visual system: changing views for a model of experience-dependent plasticity, *Neuron*, 2007, **56**: 312–326.

[33] KL Johnson *et al.*, Developmental plasticity in the human auditory brainstem, *Journal of Neuroscience*, 2008, **28**(15): 4000–4007.

[34] Reviewed in P Medini and T Pizzorusso, Visual experience and plasticity of the visual cortex: a role for epigenetic mechanisms, *Frontiers in Bioscience*, 2008, **13**: 3000–3008.

[35] PM Sonksen and N Dale, Visual impairment in infancy: impact on neurodevelopmental and neurobiological processes, *Developmental Medicine and Child Neurology*, 2002, **44**(11): 782–791.

[36] For a general introduction see KM Murphy, P Travers and M Walport, *Janeway's Immunobiology*, 7th Edition, Garland Science, 2007.

[37] MMA Fernando *et al.*, Defining the role of the MHC in autoimmunity: a review and pooled analysis, *PLoS Genetics*, 2008, **4**(4): e1000024.

As of February 2009, the IMGT/HLA database of the International Immunogenetics program (http://www.ebi.ac.uk/imgt/hla/) lists 3,371 class I and II alleles. These alleles appear to be fairly evenly distributed across the world, although there are distinct population differences.[38]

Sequence variations in MHC II can have complex consequences, including inappropriate immune reactions, as in multiple sclerosis (MS), type I diabetes, and coeliac's disease.[39] In an example of complex immune system-environment interactions, a Vitamin D response element was recently found within the promoter of the *HLA-DRB1* gene in the MHC II region associated with MS, suggesting a potential link between Vitamin D deficiency and MS.[40]

Genes within the class III MHC region determine the complement repertoire of innate immunity. Widespread variations in gene dosage and length within this region have been linked to variations in innate immune responses to pathogens.[41]

Second, the genomic sequences of T cell receptors (TCRs) are rearranged into a different sequence in each T cell. Because of the very large number of possible ways of TCR gene rearrangements, the probability that even monozygotic twins will have identical naïve T cell receptors is negligibly small.[42] In addition to rearrangements, antibodies generated by B cells are subject to somatic hypermutation during B cell maturation. Thus, for both B cells and T cells small initial differences in repertoire can be amplified by subsequent events into substantial immunologic differences, even among monozygotic twins.[43] Apart from B and T cell receptors, DNA sequence variations in other immune-related genes are

---

[38] F Prugnolle *et al.*, Pathogen-driven selection and worldwide HLA class I diversity, *Current Biology*, **15**: 1022–1027.

[39] Z Hovhannisyan *et al.*, The role of HLA-DQ8 b57 polymorphism in the anti-gluten T-cell response in coeliac disease, *Nature*, 2008, **456**: 534–538.

[40] SV Ramagopalan *et al.*, Expression of the multiple sclerosis-associated MHC class II allele HLA-DRB1*1501 is regulated by vitamin D, *PLoS Genetics*, 2009, **5**(2): e1000369.

[41] Y Yang *et al.*, Diversity in intrinsic strengths of the human complement system: serum c4 protein concentrations correlate with c4 gene size and polygenic variations, hemolytic activities, and Body Mass Index, *Journal of Immunology*, 2003, **171**: 2734–2745; EK Chung *et al.*, Genetic sophistication of human complement components C4A and C4B and RP-C4-CYP21-TNX (RCCX) modules in the Major Histocompatibility Complex, *American Journal of Human Genetics*, 2002, **71**: 823–837.

[42] T Petteri Arstila *et al.*, A direct estimate of the human αβ T cell receptor diversity, *Science*, 1999, **286**: 958–961; C Keşmir, *et al.*, Diversity of human αβ T cell receptors, *Science*, 2000, **288**: 1135.

[43] J Monteiro *et al.*, Variability in CD8+ T-cell oligoclonality patterns in monozygotic twins, *Annals of the New York Academy of Sciences*, 1995, **756**: 96–98.

also well known to underlie differences in immune response (e.g. adverse reactions to vaccines[44]).

Memory B cell and T cells of the adaptive immune response are the third cause of immunological diversity. Individuals growing up in different environments will be challenged by different pathogens and acquire distinct adaptive immunity repertoires. For example, allergies can arise from differences in childhood experiences such as the availability of mother's milk[45] and exposure to parasites and viruses[46] (the "hygiene hypothesis").

Finally, the cells and organs of both innate and adaptive immunity are subject to many of the same environmental insults as any other part of the body. Indeed, since immune cells are present throughout the body, it may be argued that the immune system is particularly susceptible to both organ-specific and general environmental insults. For example, high levels of the stress hormone cortisol have been shown to result in lowered telomerase activity in T lymphocytes during both primary and secondary stimulation.[47] T lymphocytes are normally able to up-regulate telomerase when they proliferate in response to antigen. Loss of telomerase activity leads to telomere shortening, which in turn leads to senescence. Thus, stress may damage the immune system in a lasting manner.

Figure 4.2 shows the distribution of the concentration of C-reactive protein (a marker for acute inflammation) in the blood of 143 healthy, middle-aged blood donors.[48] Note how the distribution is distinctly long-tailed and right-skewed. The individuals in the right-hand half of this distribution clearly have a different inflammatory profile from the individuals in the left-hand half.

Inter-individual immune system variability may also influence individuals in complex and indirect ways. For example, some preliminary studies have suggested that the choice of

---

[44] See for example DM Reif *et al.*, Genetic basis for adverse events after smallpox vaccination, *Journal of Infectious Diseases*, 2008, **198**(1): 16–22.

[45] V Verhasselt *et al.*, Breast milk-mediated transfer of an antigen induces tolerance and protection from allergic asthma, *Nature Medicine*, 2008, **14**(2): 170–175.

[46] JE Gern and WW Busse, Relationship of viral infections to wheezing illnesses and asthma, *Nature Reviews Immunology*, 2002, **2**: 132–138.

[47] J Choi, SR Fauce and RB Effros, Reduced telomerase activity in human T lymphocytes exposed to cortisol, *Brain, Behavior, and Immunity*, 2008, **22**: 600–605.

[48] Figure from EM Macy, TE Hayes and RP Tracy, Variability in the measurement of C-reactive protein in healthy subjects: implications for reference intervals and epidemiological applications, *Clinical Chemistry*, 1997, **43**(1): 52–58. Copyright 1997 by American Association for Clinical Chemistry, Inc. Reproduced with permission.

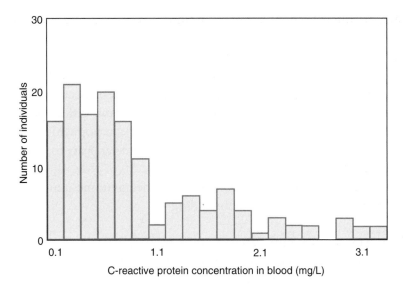

Figure 4.2: The blood concentration of C-reactive protein in 143 healthy adults.

partners may be influenced by an individual's immune repertoire, sensed through body odors and other cues.[49]

The above discussions illustrate how the immune repertoires of individuals tend to differ from the start, and become increasingly different over time. The examples provided are illustrative, not exhaustive. Moreover, inter-individual immune system differences are affected by and affect other physiological systems. For instance, the immune system is tightly coupled to the digestive system,[50] which is itself highly individualized.

The immune system in the gut (part of the mucosal immune system) is spatially segregated from the rest of the immune system. It performs the specialized functions of tolerating food antigens and commensal bacteria while detecting pathogens using a distinctive repertoire of lymphocytes. While the mechanisms of cooperation between the digestive and

---

[49] Reviewed in SC Roberts and AC Little, Good genes, complementary genes and human mate preferences, *Genetica*, 2008, **132**: 309–321. See also M Gallagher *et al.*, Analyses of volatile organic compounds from human skin, British *Journal of Dermatology*, 2008, **159**(4): 780–791.

[50] See for example D Bouskra *et al.*, Lymphoid tissue genesis induced by commensals through NOD1 regulates intestinal homeostasis, *Nature*, 2008, **456**(7221): 507–510.

immune system remain obscure,[51] both the innate and adaptive immune systems are involved,[52] and dysregulated interactions result in inflammatory bowel diseases such Crohn's.[53]

As discussed briefly in Chapter 1, commensal bacteria outnumber human cells in the body. Different mixtures of over 500 types of commensal bacteria occupy the skin, the gastrointestinal tract and the various mucosal tissues.[54] Significant differences in microbiome composition have been found between Chinese and American populations.[55] At present the cause is not clear, but the composition of the gut microbiome appears to be diet-dependent,[56] and may be responsible for some cases of obesity.[57]

Further characterization of the human microbial metagenome genome is currently the focus of a number of projects worldwide, including:

- The EU MetaHIT project (http://www.metahit.eu/metahit/)
- The Japanese Human Meta Genome project (http://www.metagenome.jp/)
- The US Human Microbiome Project[58] (http://nihroadmap.nih.gov/hmp/)
- And the Human Oral Microbiome Database (http://www.homd.org/)

Additional large-scale projects are in various stages of development in Australia, Canada, China, and France.

In addition to direct interactions between the digestive and immune systems, the enteric nervous system regulates and responds to both the digestive system and the gastrointestinal

---

[51] Reviewed in D Kelly, S Conway and R Aminov, Commensal gut bacteria: mechanisms of immune modulation, *Trends in Immunology*, 2005, **26**(6): 326–333.

[52] E Slack *et al.*, Innate and adaptive immunity cooperate flexibly to maintain host-microbiota mutualism, *Science*, 2009, **325**: 617–620.

[53] RJ Adams *et al.*, IgG antibodies against common gut bacteria are more diagnostic for Crohn's disease than IgG against mannan or flagellin, *American Journal of Gastroenterology*, 2008, **103**: 386–396.

[54] Reviewed in L Dethlefsen1, M McFall-Ngai and DA Relman, An ecological and evolutionary perspective on human–microbe mutualism and disease, *Nature*, 2007, **449**: 811–818.

[55] M Li *et al.*, Symbiotic gut microbes modulate human metabolic phenotypes, *Proceedings of the National Academy of Sciences of the USA*, 2008, **105**(6): 2117–2122.

[56] RE Ley *et al.*, Evolution of mammals and their gut microbes, *Science*, 2008, **320**(1647): 1647–1651.

[57] RE Ley *et al.*, Microbial ecology: human gut microbes associated with obesity, *Nature*, 2006, **444**: 1022–1023; PJ Turnbaugh *et al.*, An obesity-associated gut microbiome with increased capacity for energy harvest, *Nature*, 2006, **444**: 1027–1031.

[58] PJ Turnbaugh *et al.*, The human microbiome project, *Nature*, 2007, **449**: 804–810.

lymphoid tissue.[59] In this way, digestion, immunity and the nervous system are inextricably intertwined. Thus disorders and treatments addressing one of these systems can affect the other two. For instance, antibiotics (taken to aid the immune system by fighting off pathogenic bacteria) can reduce the diversity of commensal gut microbiota, which in turn can lead to proliferation of pathogenic gut bacteria and antibiotic-associated diarrhea.[60] In terms of personalized medicine, such tight linkage between three highly individualized systems underlines the need for comprehensive characterization of a person's biochemical status prior to diagnosis.

Transient Physiological Effects

In addition to the various long-term differences between individuals discussed above, various physiological factors lead to further diversity on shorter timescales. Among these factors are metabolic changes associated with eating habits, diurnal changes associated with the awake/sleep cycle, monthly hormonal changes in pre-menopausal women, and seasonal health effects associated with changes in temperature, sunlight, humidity, etc. We will briefly review a few examples below.

The diurnal cycle is well known to affect transcription levels of large numbers of genes. For example, in the mouse prefrontal cortex, approximately 10% of transcribed genes were found to oscillate in expression — up to five-fold — over 24 hours.[61] In the mouse liver, over 1,200 genes exhibited circadian expression changes, with peak to trough ratios of up to ten-fold.[62] Even the release of hematopoeitic stem cells from the bone marrow into the blood stream turns out to be regulated by circadian rhythms (peak to trough ratio ~2–3 fold). Remarkably, this diurnal pattern appears to be under direct control of the central nervous system through secretion of neurotransmitters that down-regulate the transcription of the chemokine CXCL12 in a specific sub-population of the bone marrow stromal cells.[63]

---

[59] S Ben-Horin and Y Chowers, Neuroimmunology of the gut: physiology, pathology, and pharmacology, *Current Opinion in Pharmacology*, 2008, **8**: 1–6.

[60] JY Chang *et al.*, Decreased diversity of the fecal microbiome in recurrent clostridium difficile–associated diarrhea, *Journal of Infectious Diseases*, 2008, **197**: 435–438.

[61] S Yang *et al.*, Genome-wide expression profiling and bioinformatics analysis of diurnally regulated genes in the mouse prefrontal cortex, *Genome Biology*, 2007, **8**: R247.

[62] K Oishi *et al.*, Genome-wide expression analysis of mouse liver reveals *Clock*-regulated circadian output genes, *Journal of Biological Chemistry*, 2003, **278**(42): 41519–41527.

[63] S Mendez-Ferrer *et al.*, Haematopoietic stem cell release is regulated by circadian oscillations, *Nature*, 2008, **452**: 442–448.

In summary, the expression levels of many genes, and other physiological biomarkers, varies by two-fold or more during a typical 24-hour period. We will return to this issue when we discuss measurements of health status in individuals in Chapter 7.

We noted in the section on epimutations that a diet high in methyl groups has been shown to change DNA methylation patterns (and hence coat coloring) in *agouti* mice. Many foods and drinks have toxic effects when consumed in unusually high amounts. For example, chronic alcohol abuse results in liver damage, in part due to transcriptional down-regulation of a gene (*SIRT1*) required for mitochondrial function.[64]

Interestingly, the adverse effects of chronic alcohol over-consumption are not limited to the liver. For example, in the frontal cortex, alcohol abuse leads to an increase in the levels of β-catenin protein, a substrate for GSK3β in the Wnt signaling pathway[65] (which is implicated in cancer and Alzheimer's disease). As with many other potentially toxic materials, the effect of alcohol on health seems to be biphasic: at high doses alcohol has severe adverse effects, while at low doses it appears to be beneficial. Regular consumption of small amounts of alcohol is associated with improved angiogenesis[66] and cardiovascular health.

In addition to dietary composition, changes in an individual's overall metabolic rate also appear to have complex effects on health. For example, it has been known for some time that caloric restriction, but not exercise (increased caloric expenditure), can impede carcinogenesis[67] in mice.

In a recent study, skin gene expression responses to a chemical carcinogen were measured in mice subjected to dietary caloric restriction or exercise. The responses of the two groups to carcinogens were radically different. In the diet group, 383 genes responded differently from controls (>1.5-fold change) after exposure to the carcinogen. In the exercise group, 107 genes responded differently from controls. Only two of the differentially responding genes were shared between the two groups. Thus, according to this study, reduced caloric uptake and increased caloric expenditure through exercise both affect responses to a carcinogen, but in completely different ways.

[64] CS Lieber *et al.*, Alcohol alters hepatic FoxO1, p53, and mitochondrial SIRT5 deacetylation function, *Biochemical and Biophysical Research Communications*, 2008, **373**: 246–252.
[65] AM Al-Housseini *et al.*, Upregulation of beta-catenin levels in superior frontal cortex of chronic alcoholics, *Alcoholism, Clinical and Experimental Research*, 2008, **32**(6): 1080–1090.
[66] D Morrow *et al.*, Ethanol stimulates endothelial cell angiogenic activity via a Notch- and angiopoietin-1-dependent pathway, *Cardiovascular Research*, 2008, **79**(2): 313–321.
[67] See for example CA Gillette *et al.*, Energy availability and mammary carcinogenesis: effects of calorie restriction and exercise, *Carcinogenesis*, 1997, **18**(6): 1183–1188.

Factors that affect health need not be physical. The idea that mental and emotional states can affect our health is centuries old.[68] In many cases, a correlation between mental/emotional state and physical health can be observed, but the reason for the correlation is not clear. For example, people with a history of severe depression in earlier years are more likely to develop Alzheimer's in old age,[69] but the cause of the correlation is not known.

In recent years, a number of studies have identified molecular mechanisms that mediate mind–body effects in specific cases. For example, social isolation is known to adversely affect health outcomes.[70] A recent study of social isolation (loneliness) found more than 200 differentially regulated genes (>30% difference in mean, 10% false discovery rate) in the leukocytes of lonely versus well-connected individuals.[71] A large number of the differentially regulated genes were found to be associated with increased inflammatory processes.

Social stress usually increases circulating levels of the stress hormone cortisol,[72] which can suppress inflammatory and immune responses. However, in the leukocytes of lonely individuals, glucocorticoid receptor levels (and hence transduction of the cortisol signal) are down-regulated, while the pro-inflammatory NF-κB transcriptional regulators are up-regulated.[71] Thus, the feeling of loneliness can lead to an increased risk of pro-inflammatory diseases such at atherosclerosis, and stress-reducing exercises such as yoga, tai chi, and meditation might confer molecular as well as physiological and psychological health benefits.

It should be noted that in addition to the above sources of variation within and among individuals, there appear to be more complex, perhaps random, changes in health status markers with unknown timescale. For example, figure 4.3[48] shows measurements of the blood concentration of C-reactive protein (CRP) in 26 apparently healthy volunteers over a six-month period (measured every three weeks).

Some of the outliers in the above figure (e.g. the highest readings in subjects 6, 15 and 26) may indicate illness (assuming a CRP concentration cut-off of ~10 mg/L) or other systematic effects. Excluding these outliers, there is still an approximate two-fold variation in CRP levels for most subjects.

---

[68] An excellent historical review is given in *The Cure Within: A History of Mind–Body Medicine*, by Anne Harrington, WW Norton and Company, 2008.

[69] MI Geerlings *et al.*, History of depression, depressive symptoms, and medial temporal lobe atrophy and the risk of Alzheimer disease, *Neurology*, 2008, **70**: 1258–1264.

[70] JT Cacioppo and LC Hawkley, Social isolation and health, with an emphasis on underlying mechanisms, *Perspectives in Biology and Medicine*, 2003, **46**(3): S39–S52.

[71] SW Cole *et al.*, Social regulation of gene expression in human leukocytes, *Genome Biology*, 2007, **8**: R189.

[72] EK Adam *et al.*, Day-to-day dynamics of experience — cortisol associations in a population-based sample of older adults, *PNAS*, 2006, **103**(45): 17058–17063.

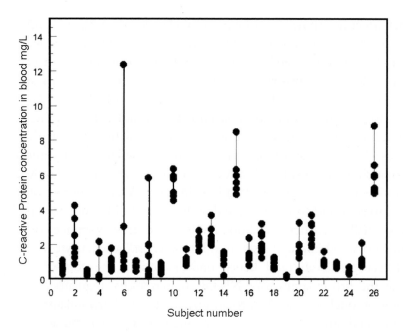

Figure 4.3:   Variation over time of blood concentrations of C-reative protein in 26 healthy adults.

Cumulative Effects

Many factors have a cumulative effect on health. Age in general, and past history of exposures in particular tend to be highly predictive of health status. For example, the causes of thrombi change dramatically with age. As illustrated in figure 4.4, the percentage of cases due to stable plaque increases with age, while the proportion of acute thrombi observed decreases with age.[73]

Past episodes of chronic illness can affect future health both directly (e.g. disability) and also indirectly. For example, chronic infections have been shown to increase susceptibility to coronary heart disease. In response to injury or pathogens, innate immune cells release inflammatory cytokines such as IL-6 and TNFα, which up-regulate the synthesis of C-reactive protein and other acute-response reactants by the liver. In a typical population, the serum concentration of C-reactive protein varies by as much as 1,000-fold between

---

[73] Figure reproduced from RP Tracy, Deep phenotyping: characterizing populations in the era of genomics and systems biology, *Current Opinion in Lipidology*, 2008, **19**: 151–157. Reprinted with permission.

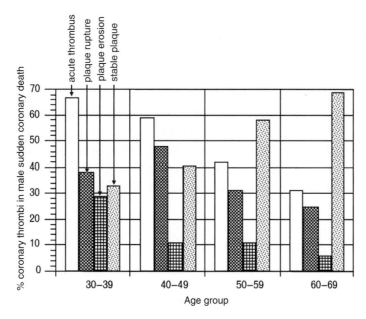

Figure 4.4: Changes in causes of thrombi with age.

individuals.[74] Much of this variation can be explained by a past history of chronic infections[74] (other contributors are smoking and obesity), and chronically high levels of C-reactive protein are strongly correlated with higher levels of coronary heart disease.

Viral infections are also implicated in the development of type 1 diabetes. In particular, certain mutations in a gene that activates the interferon response following viral infection have been shown to lower the risk of acquiring type 1 diabetes by about 50%.[75]

As noted earlier, in placental mammals environmental effects can be transmitted to growing embryos via the mother, and sometimes via effects on the paternal sperm. For example, drugs taken by pregnant women (and in certain cases, by fathers just before conception) can have damaging effects on embryonic development.[76] Illness in the mother can also affect offspring development via the uterine environment. A study comparing siblings before and

---

[74] MA Mendall *et al.*, C-reactive protein and its relation to cardiovascular risk factors: a population based cross sectional study, *British Medical Journal*, 1996, **312**: 1061–1065.

[75] S Nejentsev *et al.*, Rare variants of IFIH1, a gene implicated in antiviral responses, protect against type 1 diabetes, *Science*, 2009, **324**: 387–389; M von Herrath, A virus–gene collaboration, *Nature*, 2009, **459**: 518–519.

[76] Summarized in P Vallance, Drugs and the fetus, *British Medical Journal*, 1996, **312**: 1053–1054.

after the onset of type 2 diabetes in the mother found that children born after the mother has developed diabetes have a much higher chance of developing type 2 diabetes. Diabetes status of the father had no effect, suggesting that the effect is probably mediated through the uterine environment.[77]

Relatively benign environmental factors such as caloric or protein restriction can also affect embryonic development. Such effects often extend not only to the adult life of the offspring of affected mothers, but also to second-generation offspring because germline cells are formed during embryonic development.[78] Remarkably, over-abundance of food during a paternal grandparent's pre-puberty ("slow growth") period has also been shown to correlate with increases in cardiovascular disease in grandchildren.[79]

[77] D Dabelea *et al.*, Intrauterine exposure to diabetes conveys risks for Type 2 Diabetes and obesity, *Diabetes*, 2000, **49**: 2208–2211.

[78] Reviewed in GC Burdge *et al.*, Epigenetic regulation of transcription: a mechanism for inducing variations in phenotype (fetal programming) by differences in nutrition during early life? *British Journal of Nutrition*, 2007, **97**(6): 1036–1046.

[79] G Kaati, LO Bygren and S Edvinsson, Cardiovascular and diabetes mortality determined by nutrition during parents' and grandparents' slow growth period, *European Journal of Human Genetics*, 2002, **10**: 682–688.

CHAPTER 5

The Impact of Inter-individual Biochemical Differences on Health Outcomes

In Chapter 2, we noted briefly that although cellular pathways have traditionally been portrayed as linear cascades, they are in reality highly non-linear, both in terms of sequential dynamics, and also in terms of the inter-pathway interactions that coordinate the various aspects of cellular function (e.g. differentiation, metabolism, response to changes in extra-cellular conditions). Chapters 3 and 4 highlighted the many genetic and non-genetic ways in which any two individuals are likely to differ in abundances and states of gene products.

This chapter looks in more detail at the extent to which molecular differences among individuals may perturb cellular behavior, organ function, organism health, and responses to treatment. We survey the impact of biochemical systems complexity on organ health and function, and conclude that accurate prediction of the effects of some genetic and epigenetic inter-individual differences will require mechanistic modeling of the affected molecular systems.

Hierarchical System Modularity and Functional Building Blocks

The complexity of the human body makes it very difficult to predict how it will respond to perturbations in its make-up (genetic perturbations) and its operating environment (environmental and life-history exposures). One way to address this problem is to exploit the modular and hierarchical organization of the body.

A functional module is any set of interacting components that performs a specific function. The components of a functional module may be distributed spatially; they may interact with each other and with components of other modules briefly or over extended periods; and they may be reused as components of different modules in different contexts. Moreover, the same type of function (e.g. an on/off switch) may be realized using different components in different contexts. What distinguishes a functional module is that the interactions among its components perform a specific and well-defined function (e.g. digestion).

Human physiology can be modeled as a hierarchy of functional modules operating at the organism, organ, cellular, and pathway levels. At the largest scale (top of the hierarchy), we can view an individual's physiology as arising from the interactions of the circulatory system, the immune system, the digestive system, the nervous system, etc. Next, each of these physiological systems (functional modules) can be sub-divided into smaller functional modules.

Taking the immune system as an example, we may divide immunity into innate and adaptive sub-systems (aka functional modules), each of which can in turn be divided into smaller sub-systems such as humoral immunity and cell-mediated immunity.

Near the bottom of the body's functional hierarchy (at the smallest scale of interest to us here), individual RNA and protein molecules are encoded in DNA as sets of sequence motifs that specify the interacting structural and functional components of gene products. For example, the genomic encoding of a protein molecule may specify multiple binding domains, spatial localization signals, and chemical modification sites within the context of the protein's 3D structure. It is at this level that DNA sequence variations exert their influence.

At the next level up, gene products biochemically interact with each other in intra- and inter-cellular pathways that perform specific cellular functions. It is at this level that the effects of DNA sequence variation translate into possible variations in physiology. In a similar vein, environmental factors may also modulate the function of individual molecular species and the pathways they take part in. Perturbed pathway functions may then percolate up the hierarchy to manifest as dysregulated organ function and disease.

In the rest of this chapter, we will briefly review some examples of the ways in which changes in the abundance or interaction characteristics of gene products can affect the behavior of functional modules at the intra- and extra-cellular levels. We will then consider some of the practical difficulties of predicting the effects of novel perturbations on health status.

The Dynamic Characteristics of Molecular Interaction Modules Determine the Impact of Genetic and Environmental Factors[1]

Functional building blocks in molecular networks. A consensus view emerging from systems analyses of diverse pathways and organisms is that biochemical interaction networks within cells are organized in a highly modular fashion, with types of functional modules

---

[1] For more in-depth discussion of topics in this section, see U Alon, *An Introduction to Systems Biology: Design Principles of Biological Circuits*, CRC Press, 2006; H Bolouri, *Computational Modeling of Gene Regulatory Networks: A Primer*, Imperial College Press, 2008.

(e.g. on/off switches) being used repeatedly as "building blocks" (in the style of LEGO®
blocks).

A wide variety of functional building blocks have been proposed.[2] For illustrative purposes, we will focus here on examples from the simplest functional class: on/off switches.
However, it is important to note that there are many additional types of molecular network
building blocks, some of which may be more structurally and functionally complex than the
simple examples reviewed here.[3]

Several different frequently occurring molecular network building blocks can act as
switches. A few examples are presented below. As we shall see, different types of perturbation
in different components of each module can result in diverse functional effects. Some
perturbations can have dramatic consequences, while others may produce little or no functional effect.

All-or-nothing switches. A form of push–pull arrangement found in most signal
transduction pathways can confer an all-or-nothing, switch-like response to ligands. Two
common forms of this functional property, called "ultra-sensitivity",[4] are: (1) phosphoryla-
tion-dephosphorylation cascades,[5] and (2) active switching between transcriptional
repression and transcriptional activation.[6]

An illustrative example of a protein modification cascade is presented in the right-hand
panel of figure 5.1. The molecular species A and B represent enzymes that become activated
upon covalent modification (as in the MAP Kinase pathway[5]). The active form is indicated
by*. Each stage in the cascade acts as an amplifier of the changes in the active enzyme level
of the preceding stage. Cascading multiple stages can result in sharp, threshold-like responses
to a signal (e.g. a ligand).

---

[2] For reviews, see JJ Tyson, KC Chen and B Novak, Sniffers, buzzers, toggles and blinkers: dynamics of regula-
tory and signaling pathways in the cell, *Current Opinion in Cell Biology*, 2003, **15**: 221–231; W Longabaugh and
H Bolouri, Understanding the dynamic behavior of genetic regulatory networks by functional decomposition,
Current Genomics, 2006, 7(6): 333–341.

[3] See for example, W Ma *et al.*, Design principles of biochemical adaptation networks: enumerating solutions in
network space, *Cell*, 2009, to appear.

[4] A Goldbeter and DE Koshland, An amplified sensitivity arising from covalent modification in biological systems,
Proceedings of the National Academy of Sciences of the USA, 1981, **78**(11): 6840–6844.

[5] C-YF Huang and JE Ferrell, Ultrasensitivity in the mitogen-activated protein kinase cascade, *Proceedings of the
National Academy of the USA*, 1996, **93**: 10078–10083.

[6] S Barolo and JW Posakony, Three habits of highly effective signaling pathways: principles of transcriptional
control by developmental cell signaling, *Genes and Development*, 2002, **16**: 1167–1181.

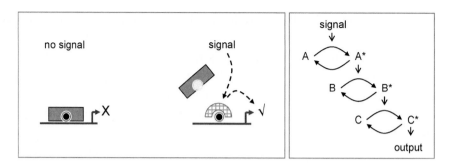

Figure 5.1:   Two types of all-or-nothing biochemical switches.

Seven common inter-cellular signaling pathways — Wnt, TGF-β, Hedgehog (Hh), receptor tyrosine kinase (RTK), nuclear receptor, Jak/STAT, and Notch — employ another mechanism for sharp switch-like responses to signals. They respond to signals by switching their downstream target genes from a state of enforced transcriptional repression to one of activation by swapping regulatory co-factors. For example, in the canonical Wnt signaling system, ligand activity results in translocation of β-catenin to the nucleus where it displaces Groucho repressor molecules bound to TCF-LEF transcription factors and activates transcription.

The left panel of figure 5.1 illustrates the above behavior schematically. Here, the solid disk represents TCF/LEF. The gray box represents Groucho, which is bound to TCF/LEF in the absence of Wnt signaling, and represses transcriptional activity. The hatched crescent symbolizes signal-activated β-catenin, which has displaced Groucho from its complex with TCF/LEF, and enabled TCF/LEF to act as a transcriptional enhancer.

The schematic in the left panel of figure 5.2 shows how sharp, threshold-like switch responses can either filter out or amplify changes in their input signal. Signal level variations within the two shaded regions have very little effect on the switch output. In contrast, small changes in the signal level within the middle region can produce large changes in the output. In particular, any factors causing a signal level shift from one shaded region to the other will maximally change the output level.

The right-hand panel in figure 5.2 illustrates three characteristic ways in which genetic or environmental factors can modify the behavior of an all-or-nothing switch. The dashed curve indicates the unperturbed (wild type) behavior of the switch. The left-most curve illustrates a perturbation causing a change in the switching threshold. In the example shown, the perturbed switch will turn on at a lower input activity level than in the wild type.

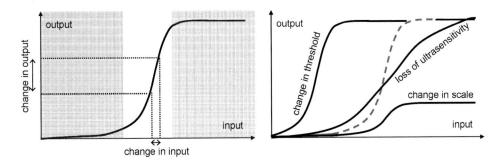

Figure 5.2:   Possible healthy and dysregulated behaviors of all-or-nothing switches.

A signaling pathway with this type of perturbation will tend to be active more often than the wild type. Not shown is the complementary situation in which the threshold is shifted upwards, resulting in an under-responsive signaling pathway.

Two other types of functional perturbation are shown: loss of threshold-like behavior (curve with lowered slope), and reduced response magnitude (right-most curve). Loss of ultra-sensitivity changes the function of an all-or-nothing switch to a graded response, which may result in inappropriate activation/repression of downstream modules. A perturbation that leads to a reduced response magnitude may cause the switch to appear as though it is in the off-state at all times.

To summarize, the response of downstream processes to a mutation or environmental perturbation in upstream signaling may be varied.

The two all-or-nothing switching mechanisms presented above are examples of single-threshold switches. Another mechanism that can produce threshold-like responses to changes in cellular conditions is compartmentalization. In compartmental systems, spatial segregation of reactants/products can enhance small differences in input conditions while blocking homogenizing processes.[7]

Some systems with positive (self-reinforcing) feedback also act as switches, but in this case, the switches can have separate on and off thresholds, as discussed below.

Two-threshold switches. Combinations of positive and negative feedback loops confer stable steady states to many cellular responses. Positive feedback loops reinforce change, and are necessary for systems with more than one stable steady state. Negative feedback

---

[7] See for example MA Daniels *et al.*, Thymic selection threshold defined by compartmentalization of Ras/MAPK signaling, *Nature*, 2006, **444**: 724–729.

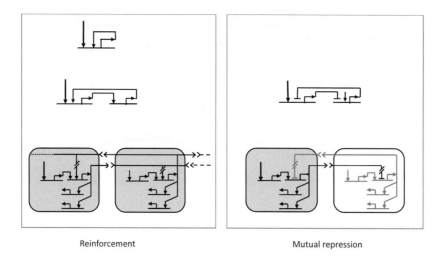

Reinforcement   Mutual repression

Figure 5.3:   Examples of positive feedback gene regulatory building blocks.

loops counteract change, and — among other things — can return a system to its set-point after a perturbation (homeostatic stability) or control the switching threshold between two states.

Figure 5.3 shows some examples of transcriptional positive feedback. The examples in the left panel all involve loops of activating (positive) interactions. The examples on the right all involve repression of a repressor (gene symbols as defined earlier; lines ending in arrowheads represent activating inputs; lines ending in bars represent repressive inputs; rounded rectangles represent cells and the double chevron symbol is used to denote a ligand-receptor interaction).

All five of the circuits shown above behave as on-off switches with different turning-on and turning-off thresholds. But the inter-cellular networks perform an important additional function: coordination of gene expression among communicating cells. The mutual repression scenario (right panel, bottom) ensures that adjacent cells express complementary sets of genes. It can support cell-type specification at tissue boundaries during embryonic development.[8]

Inter-cellular co-activation (left panel, bottom) can take a number of forms. In the example shown a ligand-encoding gene is itself a transcriptional target of its ligand's signaling

---

[8] J Jensen, Gene regulatory factors in pancreatic development, *Developmental Dynamics*, 2004, **229**: 176–200.

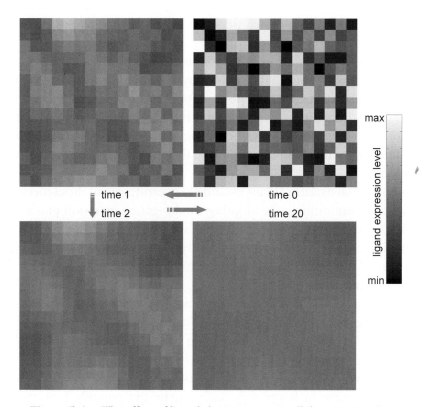

Figure 5.4: The effect of ligand sharing on inter-cellular co-activation.

pathway. In such a system, the same set of genes is activated in adjacent cells. Ligand-sharing among adjacent cells ensures that no cell is left behind: all cells within a tissue are triggered to activate the signal-mediated positive feedback loop. In this way, a noisy initial activation pattern can result in the homogenous activation of a group of adjacent cells.[9]

We will explore the behavior resulting from positive feedback loops in a moment. Figure 5.4 illustrates the effect of ligand sharing on inter-cellular co-activating genes. The top-right panel shows a 15 × 15 grid of cells in which the expression level of the ligand-producing gene in each cell has been randomized (see scale bar at right). The other three panels show

---

[9] HJ Standley, AM Zorn and JB Gurdon, eFGF and its mode of action in the community effect during Xenopus myogenesis, *Development*, 2001, **128**: 1347–1357; see also H Bolouri and EH Davidson, Modeling transcriptional regulatory networks, *Bioessays*, 2002, **24**(12): 1118–1129.

the effect of ligand sharing at three consecutive time points (anticlockwise from the top right).[10]

Each cell can sense equal amounts of ligands emanating from itself and its immediate four neighbors. The total amount of ligand sensed by a cell determines the expression of the ligand gene in that cell at the next time point. Note how, over time, the expression level of the ligand gene (and therefore its downstream targets) is equalized in all the cells. In this example, ligand sharing is assumed to occur only among cells with shared boundaries. Longer-range diffusion of the ligand would make the equalizing effect of ligand sharing even more pronounced.

Thus, the above co-activation circuit essentially overrides small differences between cells, and results in all communicating cells adopting similar gene expression patterns. In contrast, inter-cellular mutual exclusion ensures that even small differences between communicating cells are amplified into distinct differential gene expression patterns. Within a single cell, co-activation and mutual inhibition can coordinate the activity of complementary and opposing intra-cellular processes respectively.

Both types of positive feedback confer uniform, all-or-nothing responses to input signals and filter out small deviations in input activity. Wnt signaling is thought to be engaged in a co-activating positive feedback loop of this type in breast and ovarian cancers,[11] as well as in the hematopoietic stem cells of the bone marrow.[12]

To better understand the dynamic behavior of positive feedback switches, consider the simple auto-regulatory gene in figure 5.5. Such a gene can typically have two stable steady states: no expression, and full expression. These stable steady states can be visualized as energetically favorable states in a landscape of all possible system states; analogous to a ball rolling to the bottom of one of two adjacent troughs (see cartoon in the middle panel in figure 5.5).

The graph in the right-hand panel of the figure shows the steady state expression level of the gene (mRNA or protein) as a function of the level of input activity. The upper and lower solid curves represent the on and off states of the gene. Note that the output is essentially either fully on or fully off, and that the two curves overlap along the horizontal axis. This is

---

[10] For a detailed discussion of this model, see H Bolouri and EH Davidson, The gene regulatory network basis of the "community effect," and analysis of a sea urchin embryo example, *Developmental Biology*, 2009, DOI: 10.1016/j.ydbio.2009.06.007.

[11] A Bafico *et al.*, An autocrine mechanism for constitutive Wnt pathway activation in human cancer cells, *Cancer Cell*, 2004, 6: 497–506.

[12] Reviewed in FM Rattisa, C Voermans and T Reya, Wnt signaling in the stem cell niche, *Current Opinion in Hematology*, 2004, 11: 88–94.

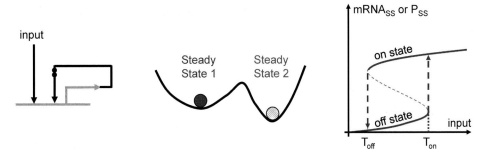

Figure 5.5: A two-threshold switch using auto-regulatory positive feedback.

because the system switches at a different threshold if we start with the gene off and try to turn it on, versus if we start with the gene on and try to turn it off (follow the dashed arrows). The thin dashed line connecting the two steady state loci is analogous to the ridge separating the two troughs in the middle panel.

Although the example illustrated here is for a single auto-regulatory gene, it can be shown that the mutual repression and co-activation circuits discussed earlier behave similarly, and this is true for both the intra- and inter-cellular versions.

Compared to the ultra-sensitive switches discussed earlier, two-threshold positive feedback switches show more resistance to small variations in their inputs (due to the distance between T_{OFF} and T_{ON}). The right-hand panel in figure 5.6 shows three simple ways (numbered) in which two-threshold switches can be affected by mutations in their components:

(1) Increased or reduced distance between thresholds (dashed curve). The result is a switch that is either difficult to flip (increased distance between thresholds), or a switch that may flip in response to noise in the input (reduced distance between thresholds).
(2) Graded instead of binary response (dotted curve). In this case, we no longer have two distinct steady states, just a quiescent state set by the input level.
(3) Reduced (or increased) fold-change in output upon switching (solid curve). In the figure, only a change in the higher steady state value is shown. The lower steady state can be similarly affected (i.e. shifted to a higher or lower magnitude).

The left panel of figure 5.6 shows a fourth type of effect. In the example shown, a lower threshold at which the positive feedback loop becomes dominant has shifted the entire response curve to the left. As a result, the on-state to off-state transition (downward arrow, crossed) now occurs only for input values below zero (shaded region). The upshot is a

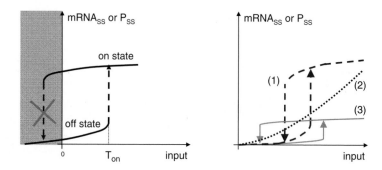

Figure 5.6:   Some possible effects of sequence variations on two-threshold switches.

bistable switch that, once activated, cannot be turned off (because it requires negative input activity levels). A complementary situation arises if the on-threshold is shifted up (to the right) so that the switch cannot turn on for physiological levels of input activity.

The behavior of two-threshold (aka hysteretic) switches illustrates an important point regarding the effect of genetic variations and environmental factors on cellular function. A partial loss or increase in input activity (e.g. a heterozygote deletion or duplication) may have little functional effect, whereas mutations that affect the feedback loop itself can have a dramatic effect.[1]

The Challenges of Establishing Cause–Effect Relationships

Distributed regulation and apparent redundancy. Because of the high degree of regulatory interactions in intra- and inter-cellular networks, genetic and epigenetic perturbations can often have complex effects. Metabolic networks are typically regulated simultaneously at many points. As a result, the enzymes that regulate metabolism are rarely "rate limiting" and a mutation in any one metabolic enzyme may not have dramatic effects on system behavior as a whole.[13] Thus, although multi-point regulation may *appear* redundant, it is important in conferring robustness to metabolic pathways.

While individual metabolic pathways are robust to some perturbations, metabolic networks often cross-regulate each other, so that a perturbation in one metabolic pathway can result in widespread perturbations in other pathways.

---

[13] DA Fell and S Thomas, Physiological control of metabolic flux: the requirement for multisite modulation, *Biochemical Journal*, 1995, **311**: 35–39.

Another form of apparent redundancy arises from the presence of multiple forms of a gene product. In humans, most proteins have multiple isoforms, resulting from alternative splicing and/or multiple versions of a gene. For example there are 19 Wnt ligands in mammals. The isoforms of a protein often have similar interaction profiles, so that when one isoform is dysfunctional, other isoforms can substitute for it.

For proteins that undergo a chain of covalent modifications (e.g. at multiple phosphorylation sites), the overall rate of the reaction is proportional to the *product* of the rates of the individual steps. Thus, two very similar proteins will undergo a multiple modification process at very different overall rates. This phenomenon, known as "kinetic proof-reading", can allow multiple isoforms of a single protein to perform very different functions in normal cells.[14]

In diseased cells, often a different isoform substitutes for a dysfunctional protein. As a result, two pathways that are normally differentially regulated (by different isoforms) in healthy conditions become co-dependent on a single isoform, with potentially cryptic effects.

Coordination of activity among multiple functional modules. Cellular processes are coordinated via multiple cross-regulatory interactions. For example, figure 5.7 shows some of the known interactions among the Wnt, TGFβ (Transforming Growth Factor beta), and cadherin pathways, which jointly regulate the processes of cellular growth, adhesion and migration.[15] In the figure, the arrows indicate activating (positive) influences and lines ending in bars indicate repressive (negative) interactions. In addition to Wnt, TGFβ, and cadherin, a few key downstream effectors are also shown to highlight the mechanisms of pathway interaction and specificity.

Canonical Wnt signaling nuclearizes β-catenin (near the bottom-right), whereas cadherin sequesters away β-catenin at the cell surface. Thus, cadherin-mediated cell-adhesion counteracts canonical Wnt signaling. Moreover, the *Slug* gene is transcriptionally activated by TCF/LEF (activated in canonical Wnt signaling) and represses *cadherin* (TCF/LEF may also directly bind to and repress the transcription of *cadherin*).

The outcome of these interactions is a mutual exclusion switch between cadherin-mediated cell adhesion and canonical Wnt signaling. However, these are not the only interactions cross-regulating Wnt and cadherin. For example, the growth factor TGFβ simultaneously

---

[14] PS Swain and ED Siggia, The role of proofreading in signal transduction specificity, *Biophysical Journal*, 2002, **82**: 2928–2933.

[15] Reviewed in WJ Nelson and R Nusse, Convergence of Wnt, β-catenin, and cadherin pathways, *Science*, 2004, **303**: 1483–1487.

Personal Genomics & Personalized Medicine

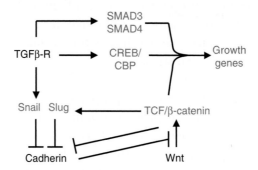

Figure 5.7: Regulatory interactions among Wnt, cadherin, and TGFβ pathways.

represses cadherin activity (via the transcriptional repressor Snail) while working synergistically with TCF/LEF to activate pro-growth genes.

Cross-regulatory interactions among "pathways" can spread the effects of genetic and epigenetic mutations. In the above example, loss of cadherin activity would not only reduce cell adhesion capacity, it would also facilitate cell proliferation (by freeing β-catenin that would otherwise be sequestered by cadherin).

Emergent behavior at the organ level. In addition to intra-cellular and intra-tissue regulation, physiological processes are also tightly cross-regulated at the organ and multi-organ levels. Indeed, traditionally more data has been available at this scale than at the intra-cellular level, and there is a rich history of integrative physiological modeling aimed at interpreting clinical data without recourse to detailed models of molecular interaction networks.[16]

Even given a detailed understanding of the pertinent intra-cellular molecular processes, it is often the case that the effects and significance of dysregulated gene expression may only become apparent at the organ level. Diseases affecting the rhythmic patterns of movement in the heart provide good examples. At the cellular level, the mechanical movements of the heart are due to complex, biochemically regulated flow of ions across cell membranes.

---

[16] See for example AC Guyton *et al.*, Systems analysis of arterial pressure regulation and hypertension, *Annals of Biomedical Engineering*, 1972, **1**: 254–281; JX Polaschek *et al.*, Using belief networks to interpret qualitative data in the ICU, *Respiratory Care*, 1993, **38**(1): 60–71.

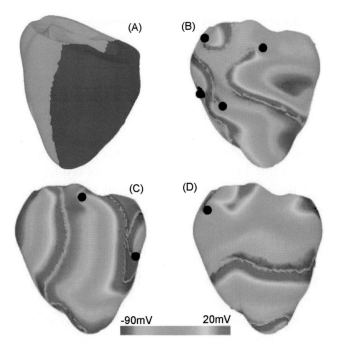

Figure 5.8: The influence of spatial cellular organization on ventricular fibrillation.

But to predict patterns of arrhythmia accurately, it is necessary to capture the three-dimensional anatomy and cellular organization of the organ.[17]

By way of an example, figure 5.8 illustrates graphically how spatial heterogeneity in cell physiology (here, variations in action potential duration) can affect patterns of ventricular fibrillation (VF).[18] In this anatomically detailed model of the rabbit heart, cardiac cells can

---

[17] For a review of models of heart physiology, see D Noble, Computational models of the heart and their use in assessing the actions of drugs, *Journal of Pharmacological Sciences*, 2008, **107**: 107–117; also, RL Winslow *et al.*, Electrophysiological modeling of cardiac ventricular function: from cell to organ, *Annual Review of Biomedical Engineering*, 2000, **2**: 119–155.

[18] These figures are reprinted with permission from H Arevalo, B Rodriguez and N Trayanova, Arrhythmogenesis in the heart: multiscale modeling of the effects of defibrillation shocks and the role of electrophysiological heterogeneity, *Chaos*, 2007, **17**: 015103. Copyright Chaos 2007, American Institute of Physics.

be divided into two distinct classes: short and long duration action potential generators,[19] as indicated by the shading in panel (A).

Panels (B–D) show simulation results three seconds after induction of VF. The shading in (B–D) has a different meaning from that in (A): it indicates the transmembrane potential distribution (see scale bar at the bottom). The model in (B) has a cell-type distribution as in (A). (C) and (D) correspond to models in which all cells behave identically. In (C), all cells produce short action potentials. In (D) all cells generate long action potentials. Note how the predictions of the two simplified models have virtually nothing in common with the more detailed model.

We noted in Chapters 2 and 4 the enormous diversity of cell types in the immune system and the microbiome of the gut. Because of the complexity of these systems as a whole, cell–cell interactions within these systems, between them, and between these two systems and the nervous system, are not well understood. However, it is already clear that complex multi-step interactions between dysregulated immune cells and host cells underlie diseases ranging from autoimmunity to epithelial tumors.[20]

In short, understanding many diseases will require not only an understanding of the affected intra-cellular biochemical pathways, but also a clearer picture of higher level interactions among cells and organs. Except in a few specific areas, such as cardiac physiology, multi-scale models spanning from molecular interactions to whole organ function are still in their infancy and the subject of intense ongoing research. Thus, for the next several years, our models of the effects of genetic and epigenetic perturbations on health are likely to remain probabilistic and correlational.

Network size. The examples presented in the preceding sections were chosen for their simplicity in order to facilitate the communication of key concepts. Most mammalian cellular processes involve interactions not among one, two or three gene products but among dozens, hundreds or thousands of them. This point is well illustrated by the existing integrative network maps describing metabolic pathways,[21] signal transduction in innate immune cells,[22] and apoptosis (programmed cell death).[23]

---

[19] In this study, to simplify the modeling process cells were "lumped" together in tissue sections of length 0.3–1 mm.

[20] KE de Visser *et al.*, Paradoxical roles of the immune system during cancer development, *Nature Reviews Cancer*, 2006, **6**: 24–37.

[21] See for example http://tinyurl.com/SigmaAldrichMetabolicChart.

[22] K Oda and H Kitano, A comprehensive map of the toll-like receptor signaling network, *Molecular Systems Biology*, 2006, **2**: 2006.0015.

[23] KW Kohn and Y Pommier, Molecular interaction map of the p53 and Mdm2 logic elements, which control the off–on switch of p53 in response to DNA damage, *Biochemical and Biophysical Research Communications*, 2005, **331**: 816–827.

The large size of mammalian intra-cellular networks, combined with the abundant cross-regulatory and feedback paths within and between networks makes prediction of the effects of genetic and epigenetic variations challenging. As a specific example, let us consider the role of the tumor suppressor gene *TP53* (encoding the protein p53) in apoptosis. On detection of DNA damage, the p53 system suspends further cell division, invokes the DNA repair system, and if repair fails, activates programmed cell death.[24]

Because *TP53* is mutated in more than 50% of tumors,[25] the p53 network is one of the most widely studied in humans. More than 50,000 journal papers have been published on p53 since it was first discovered in 1979. It even has its own biannual conference series. We therefore know much more about the regulatory interactions that determine the levels of p53 and its actions than most other cellular processes. As such, p53 control of apoptosis offers a good example of the level of complexity we may expect to encounter in well-studied human regulatory pathways.

p53 is a transcription factor that binds DNA as a tetramer via sequence-specific and DNA-structure dependent recognition domains.[26] It can act as both a transcriptional activator and also as a repressor in a complex target-dependent manner.[27] The human p53 protein consists of 393 amino acids. The latest (September 2008) issue of the p53 mutation database (http://p53.free.fr/) lists about 27,900 instances of p53 mutations.[28] About 90% of these mutations occur in the central, sequence-specific DNA-binding domain of p53. But individual cancers have distinct p53 mutation hotspots, as illustrated in figure 5.9. Note the mutation hotspots at codons 245 and 282 for gastric, but not breast cancer.[29]

Why different types of aberrant p53 DNA binding should have different frequencies in various cancers is not clear at present, but specific p53 mutations are associated with different carcinogens. For example, consumption of Aflotoxin-contaminated foods is associated

---

[24] FJ Geske *et al.*, DNA repair is activated in early stages of p53-induced apoptosis, *Cell Death and Differentiation*, 2000, **7**: 393–401.

[25] JA Royds and B Iacopetta, p53 and disease: when the guardian angel fails, *Cell Death and Differentiation*, 2006, **13**: 1017–1026; K Kurose *et al.*, Frequent somatic mutations in PTEN and TP53 are mutually exclusive in the stroma of breast carcinomas, *Nature Genetics*, 2002, **32**(3): 355–357.

[26] E Kim and W Deppert, The versatile interactions of p53 with DNA: when flexibility serves specificity, *Cell Death Differentiation*, 2006, **13**: 885–889.

[27] O Laptenko and C Prives, Transcriptional regulation by p53: one protein, many possibilities, *Cell Death and Differentiation*, 2006, **13**: 951–961.

[28] See also the International Agency for Research on Cancer (IARC) TP53 Mutation Database at http://www-p53.iarc.fr.

[29] This figure and associated data are reproduced from L Hjortsberg *et al.*, The p53 Mutation handbook 2.0, available freely from http://p53.free.fr. Figures reproduced by kind permission of Dr. Thierry Soussi.

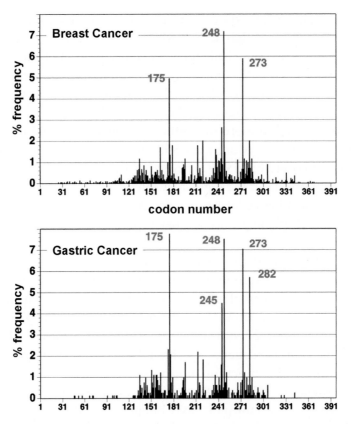

Figure 5.9: Frequencies of p53 mutations in two types of cancer.

with p53 mutations involved in liver cancer,[30] while the p53 mutations predominant in smokers are different from the p53 mutations observed in non-smokers.[31]

Intriguingly, p53 is polymorphic at codons 47 and 72 in the general human population. Minor alleles at these sites predispose carriers to poor cancer prognosis. The fact that these cancer-associated polymorphisms have not been selected against may be related to additional functions for p53 beyond control of apoptosis (e.g. senescence, inflammation).[25,32]

---

[30] F Staib *et al.*, TP53 and liver carcinogenesis, *Human Mutation*, 2003, **21**: 201–216.

[31] TM Hernandez-Boussard and P Hainaut, A specific spectrum of p53 mutations in lung cancer from smokers: review of mutations compiled in the IARC p53 database, *Environmental Health Perspectives*, 1998, **106**(7): 385–391.

[32] ME Murphy, Polymorphic variants in the p53 pathway, *Cell Death and Differentiation*, 2006, **13**: 916–920.

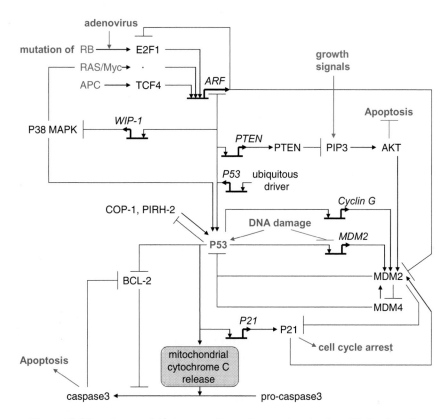

Figure 5.10: Some of the key regulatory interactions in the p53 "pathway".

The schematic figure 5.10 shows a small portion of the known interaction network for p53. Genes are shown as thick horizontal lines (representing DNA) from which a bent arrow emerges (denoting transcription/translation). Proteins are indicated with a name alone. Arrows mean activation, while lines ending in bars indicate repression. The p53 regulatory system has been described as having five parts:[33] (1) condition detection; (2) signal processing and transduction; (3) the core P53 control module; (4) response transduction; (5) effector gene batteries.

---

[33] AJ Levine, W Hu and Z Feng, The P53 pathway: what questions remain to be explored? *Cell Death and Differentiation*, 2006, **13**: 1027–1036; see also B Vogelstein, D Lane and AJ Levine, Surfing the p53 network, *Nature*, 2000, **408**: 307–310.

p53 is known to interact with around 80 other proteins.[34] To make the diagram comprehensible, many interactions are not shown here, and the first two and last two stages in the p53 system are only shown in highly simplified forms. The effects of various oncogenes and of adenoviral infections are represented here simply as causing activation of the *ARF* gene,[35] growth signals are shown as up-regulating PIP3 activity, and the downstream effectors of p53 mediated cell cycle arrest and apoptosis are summarized as acting through p21 and caspase3 respectively. Other p53-mediated processes such as DNA repair[36] are not shown.

p53 has two other family members: p63 and p73 (the numbers following the letter "p" refer to the molecular weights). Each of these proteins has multiple alternative-splicing isoforms. Some of the isoforms can act as dominant negative suppressors of p53 targets, whereas others can activate p53 target genes.[37] Mdm2, the central inhibitor of p53 activity, also requires an interacting partner (Mdm4) for full p53 inhibition.[38] All of these complexities are absent in the above figure for the sake of making the figure legible, but they must be taken into account when considering the effects of genetic and epigenetic perturbations.

Even the highly simplified network diagram in figure 5.10 is so complex as to make it very difficult to predict the effect of perturbations on its behavior (try to guess the behavior of p53 in response to irradiation before reading the next paragraph). Instead of one or two feedbacks, the core network shown includes at least a dozen interacting positive and negative feedback loops, each of which can lead to complex behaviors. Some of these loops are listed in Table 5.1.[39]

---

[34] AW Braithwaite, G Del Sal and X Lu, Some p53-binding proteins that can function as arbiters of life and death, *Cell Death and Differentiation*, 2006, **13**: 984–993.

[35] For a review of the upstream activators of p53, see MF Lavin and N Gueven, The complexity of p53 stabilization and activation, *Cell Death and Differentiation*, 2006, **13**: 941–950.

[36] SA Gatz and L Wiesmüller, p53 in recombination and repair, *Cell Death and Differentiation*, 2006, **13**: 1003–1016.

[37] F Murray-Zmijewski, DP Lane and J-C Bourdon, p53/p63/p73 isoforms: an orchestra of isoforms to harmonise cell differentiation and response to stress, *Cell Death and Differentiation*, 2006, **13**: 962–972; AI Zaika and W El-Rifai, The role of p53 protein family in gastrointestinal malignancies, *Cell Death and Differentiation*, 2006, **13**: 935–940.

[38] S Francoz *et al.*, Mdm4 and Mdm2 cooperate to inhibit p53 activity in proliferating and quiescent cells *in vivo*, *Proceedings of the National Academy of Sciences of USA*, 2006, **103**(9): 3232–3237; J-C Marine *et al.*, Keeping p53 in check: essential and synergistic functions of Mdm2 and Mdm4, *Cell Death and Differentiation*, 2006, **13**: 927–934.

[39] See also SL Harris and AJ Levine, The p53 pathway: positive and negative feedback loops, *Oncogene*, 2005, **24**: 2899–2908.

Table 5.1:   Feedback loops in the p53 network.

Number	Nodes in feedback loop	Type of feedback
1	P53 ⊣ MDM2 ⊣ P53	Negative
2	P53 → COP-1 ⊣ P53	Negative
3	P53 → PIRH-2 ⊣ P53	Negative
4	MDM4 → MDM2 ⊣ MDM4	Negative
5	P53 ⊣ ARF ⊣ MDM2 ⊣ P53	Negative
6	E2F1 → ARF ⊣ E2F1	Negative
7	P53 → WIP-1 ⊣ P38 → P53	Negative
8	P53 → P21 → MDM2 ⊣ P53	Negative
9	P53 → Cyclin G → MDM2 ⊣ P53	Negative
10	BCL-2 ⊣ Caspase 3 ⊣ BCL-2	Positive
11	P53 → PTEN ⊣ PIP3 → AKT → MDM2 ⊣ P53	Positive
12	P53 → MDM2 ⊣ MDM4 ⊣ P53	Positive

To make the analyses tractable, computational models of p53 behavior tend to simplify the already simplified picture in figure 5.10 further and consider the potential behaviors of just a few feedback loops.[40] In 10 Gy gamma-irradiated MCP-7 breast cancer cells, p53 levels have been shown to oscillate in about 60% of cells with a period of ~5.5 hours over several days.[41] In these cells, the positive feedback loop involving PTEN and AKT is not active. However, theoretical studies suggest that p53 levels are likely to oscillate in the presence of PTEN also.[42]

p53 is expressed at low levels prior to irradiation, and is localized to the cytoplasm. In the study described above, the fraction of cells with oscillating nuclear p53 levels increased with the level of irradiation, but the frequency of oscillations was largely unchanged. Repeated pulses of high nuclear p53 concentration are thus translated by downstream interactions into a switch-like apoptotic response.

These observations are consistent with two lessons from analysis of the dynamics of minimal network building blocks. Firstly, negative feedback regulation (as for p53) can generate

---

[40] Reviewed in RF Horhat, GI Mihalas and M Neamtu, The p53 network modeling — current state and future prospects, *Studies in Health Technology and Informatics*, 2008, **136**: 561–566.

[41] N Geva-Zatorsky *et al.*, Oscillations and variability in the p53 system, *Molecular Systems Biology*, 2006, **2**: 2006.0033.

[42] A Cilibreto, B Novak and JJ Tyson, Steady states and oscillations in the p53/Mdm2 network, *Cell Cycle*, 2005, **4**(3): 488–493.

lasting oscillations.[43] Secondly, p53 activation of apoptosis involves a mutual repression positive feedback loop between BCL-2 and caspase-3, which has been shown to be capable of acting as a bistable switch.[44]

Thus, the essence of how p53 mediates apoptosis can be explained in terms of interacting functional building blocks. On the other hand, understanding the manner in which all the additional feedback paths modulate this core functionality will require considerable further experimental and theoretical analysis. In particular, p53 appears to be involved in feedback loops at the inter-cellular and single-gene levels also. At the cellular level, p53 activation leads to the secretion of various molecules that reshape the local cellular environment.[33] At individual gene promoters, activated p53 appears to enhance its own ability to activate transcription by recruiting co-factors for chromatin remodeling, nucleosome repositioning, and the formation of the pre-initiation complex.[27]

It is also important to note that almost all the above experimental and computational observations relate to the behavior of (cancer) cell lines. The behavior of p53 in primary human tissues is less characterized, but significant differences may be expected. For example, normal fibroblasts taken from animals and subjected to irradiation undergo cell cycle arrest, whereas transformed fibroblasts undergo apoptosis.[33] Moreover, the p53 pathway behaves differently in different primary cell types. For example, unlike fibroblasts, irradiated primary T cells undergo apoptosis.[33] Thus, both the differentiation state of the cell and interactions with other pathways influence the output of the p53 pathway.

The p53 story provides a concrete example of the challenge facing personal genomics and personalized medicine. Suppose genetic and other assays reveal one or more perturbations in components of the core p53 regulatory system in a particular individual. We then need to establish (a) how these perturbations affect the behavior of the system in the presence of all the feedbacks discussed above, and (b) what therapeutic strategies are most likely to restore the behavior of the system to its healthy state.[45] The former will require computational analysis of the type discussed above and in Chapter 8. The latter requires an ability to quantitatively measure responses to treatment non-invasively, or using very small numbers of cells collected from the patient, as will be discussed in Chapter 7.

---

[43] N Rosenfeld *et al.*, Accurate prediction of gene feedback circuit behavior from component properties, *Molecular Systems Biology*, 2007, **3**: 143.

[44] EZ Bagci *et al.*, Bistability in apoptosis: roles of Bax, Bcl-2, and mitochondrial permeability transition pores, *Biophysical Journal*, **90**(5): 1546–1559.

[45] For example approaches, see KG Wiman, Strategies for therapeutic targeting of the p53 pathway in cancer, *Cell Death and Differentiation*, 2006, **13**: 921–926.

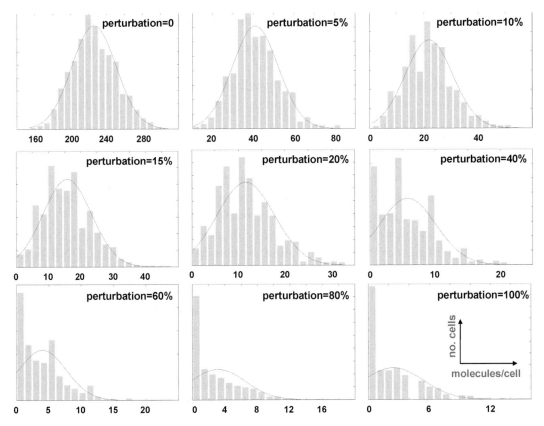

Figure 5.13:   Simulated distribution of mRNA abundance in cells downstream of dysregulated signaling.

histograms for unperturbed, signal-activated cells (top-left panel), and cells in which the signal transduction efficiency is progressively reduced by a perturbation. To be conservative, instead of a >10-fold range in wild-type mRNA variability, a conservative two-fold range is considered here (see spread along the horizontal axis in the top-left panel). Each panel shows the distribution of mRNA abundance (in molecules per cell) for a total of 1000 simulated cells.

Note how, in every case, some cells in the population are far from average. In the presence of noisy gene expression, some cells will respond to genetic and medical perturbations in a manner very different from the average. In diseases where one rogue cell may be sufficient for poor prognosis (e.g. cancer) such cell-to-cell variability can have a critical impact on treatment success rates.

Making Predictive Systems Models of Cellular and Organ Function

While the complexities and modeling challenges discussed above are considerable, they are increasingly being resolved through systems biology approaches. In essence, these approaches take advantage of three simplifying principles. Firstly, as discussed earlier, large interaction networks can often be understood in terms of functional building blocks and modules with stereotypic behaviors.

Secondly, the building block concept can be applied at multiple levels of abstraction. To predict the behavior of a system, it is usually not necessary to model every interaction at the highest possible resolution. In particular, we can model the perturbed portion of a system in the detail necessary to capture all observed phenomena, but parts of the system not directly affected by the perturbation can often be characterized at lower resolution (i.e. with less detail).

For example, transcription factors (TFs) can be modeled as being built from a collection of functional building blocks such as RNA and protein interaction domains, DNA binding domains, nuclear import/export signals, transcription activation domains, etc. We could predict the functional effect of a mutation in a TF from the domain in which it occurs. Once we have characterized the mutation in this way, we can consider its network effects in terms of the predicted functional disruption rather than molecular domain structure. Likewise, effects of perturbations on organ physiology can be modeled in terms of dysregulated functions and processes rather than individual molecular interactions.

The third simplifying principle that can help us understand cellular and organ physiology is "abstraction". Where detailed quantitative data is not available, we can build lower-resolution, more fuzzy models. For example, if we find that we are unable to construct a high-confidence mechanistic model of the effects of a mutation, we may instead be able to use population data to build a phenomenological model that associates the mutation with observed phenotypes. The phenotype association may suggest possible pathways in which the mutation is implicated, leading to a more detailed, experimentally testable model, and is a useful interim model in its own right.

This is essentially how all science proceeds. All models are approximations, and all model building is an iterative process of increasing model refinement.[57] What is different here is that we now have the technology to seek and find molecularly detailed diagnosis. In effect, we are using any and all available knowledge to guide the search for more predictive models that can identify effective interventions. We will discuss current research to predict the effects of genetic and environmental factors on cellular and organ physiology in Chapters 8 and 9.

---

[57] For a more detailed discussion of modeling concepts and approaches, see H Bolouri, *Computational Modeling of Genetic Regulatory Networks — A Primer*, Imperial College Press, 2008.

CHAPTER 6

DNA Testing and Sequencing Technologies

This chapter reviews the techniques and technologies used to identify and quantify DNA sequence variations in individuals. We briefly review established methods of testing for genetic disorders, then consider the emerging technologies for whole-genome sequencing. We note that because of its lower lifetime costs and more comprehensive information, personal whole-genome sequencing is likely to become the dominant means of assessing genetic susceptibility in the near future.

Genetic analysis can be used for a variety of medical purposes:

- **Diagnostic testing** is performed to check a potential cause of observed symptoms. For example, increasing muscular weakness, delayed developmental milestones and other symptoms in a small (<5 years old) child may point to possible Duchenne muscular dystrophy (DMD). Since DMD arises from mutations in a single gene (DMD), sequencing this single gene can confirm or reject the diagnostic hypothesis.

  DMD is a Mendelian disease, which can be detected with a highly specific, targeted genetic test. For complex (multi-factorial) diseases, it is necessary to perform diagnostic tests on multiple contributing genes. As discussed in Chapters 3 and 5, complex diseases arise from combinatorial interactions among multiple alleles and environmental factors.

- **Carrier testing** (for prospective parents) aims to identify individuals from at-risk populations who have recessive alleles for specific diseases. For example, screening for carriers of β-thalassemia in Cyprus, and for Tay-Sachs in the American Ashkenazi Jewish community has been practiced for more than two decades.[1]

  Sometimes the results of genetic carrier tests are not definitive. For example, Cystic Fibrosis (CF) is an autosomal recessive disease which is easily detected by measuring sweat

---

[1] R Schwartz Cowan, *Heredity and Hope: The Case for Genetic Screening*, Harvard University Press, 2008.

chloride levels. However, as we noted in Chapter 3, CF can arise from over 1500 different mutations in the *CFTR* gene. Targeted CF genetic tests typically only check for two to three dozen of the most common mutations. Thus a negative screening result does not guarantee that a prospective parent is not a carrier. As a result, prospective parents who are at high risk for having a child with CF sometimes choose *in vitro* fertilization and pre-implantation genetic diagnosis. If the embryo is found to have two mutated *CFTR* genes, it is not implanted in the uterus.

- **Prenatal testing** can be used for both screening and diagnostic purposes. Screening methods are usually indirect and non-invasive (e.g. ultrasound scans). More invasive diagnostic tests are performed when one of the parents is a known carrier.[2] In either case, the aim is to provide an opportunity for informed decision in cases of severe genetic disorders. Chorionic Villus Sampling (CVS) and Amniocentesis provide fetal DNA samples (from the placenta and amniotic fluid respectively). Both involve a finite risk of miscarriage and are therefore used for diagnostic purposes only. In a similar vein, Percutaneous Umbilical Blood Sampling (PUBS) is used to provide a fetal blood sample where a platelet disorder such as sickle cell anemia is suspected.

    Small amounts of fetal DNA fragments can be detected in maternal plasma. A recent study using "next-generation" high-throughput sequencing demonstrated the feasibility of detecting fetal trisomies from 7–15 milliliters of maternal blood.[3] Thus in the near future, it may be possible to test for such disorders (e.g. Down, Edwards, and Patau syndromes) from a simple maternal blood draw.

- **Newborn screening.** In most technologically developed countries newborns are routinely screened for a variety of genetic disorders. Typically, just before leaving the hospital, the baby's heel is pricked and several drops of blood are collected and placed on a filter paper which is sent to an outside laboratory for testing. The March of Dimes foundation currently recommends screening for 29 disorders.[4]

---

[2] It has been known for sometime that fetal cells and DNA are present in maternal blood, but tests using maternal blood have not yet reached clinical maturity. See for example HC Fan *et al.*, Noninvasive diagnosis of fetal aneuploidy by shotgun sequencing DNA from maternal blood, *Proceedings of the National Academy of Sciences of the USA*, 2008, **105**(42): 16266–16271.

[3] HC Fan *et al.*, Noninvasive diagnosis of fetal aneuploidy by shotgun sequencing DNA from maternal blood, *Proceedings of the National Academy of Sciences of USA*, 2008, **105**(42): 16266–16271.

[4] ***Amino acid metabolism disorders***, e.g. Phenylketonuria, Maple syrup urine disease, Homocystinuria, Citrullinemia, Argininosuccinic academia, and type I Tyrosinemia. ***Organic Acid Metabolism Disorders***, e.g. Isovaleric academia, type I Glutaric acidemia, and HMG-CoA lyase. ***Fatty Acid Oxidation Disorders***, e.g. Long Chain 3-Hydroxyacyl-CoA Dehydrogenase Deficiency, Medium Chain Acyl-CoA Dehydrogenase

- **Predictive (pre-symptomatic) testing** (susceptibility prediction) is only useful if diagnosis allows interventions that will reduce risks, or permit more effective treatment/management of symptoms after onset.[5] For complex diseases, individual alleles usually have very low penetrance, so that testing positive for any single allele is not sufficiently predictive to warrant action.[6]

  As discussed earlier, the combination of personal genomes and integrative modeling of multiple allele interactions promises to improve the accuracy with which we relate genetic and epigenetic factors to disease susceptibility, thus increasing the repertoire of diseases for which predictive testing will be beneficial.

- **Exploration and discovery science.** A cornerstone of "classical" genetics research is the identification of genotype–phenotype correlations in populations (e.g. genome-wide association studies, affected-family/linkage analysis). Another is the characterization of DNA sequence variations across geographical regions and generations (e.g. haplotype characterization, population stratification). While these studies are fundamental to genetics research, how they are designed, the analysis methods used, etc. fall beyond the scope of this book.

  Personalized medicine will benefit from the discoveries made through population studies, but we are concerned here only with the use of genetic testing and genome sequencing in individuals. These uses fall into one of the categories discussed above.

Most genetic conditions can be tested through a variety of technologies (see the discussion below). The choice of test(s) performed will depend on the specifics of the condition. To see why, consider tests for the Prader–Willi Syndrome (PWS). PWS is a complex genetic disorder that typically causes low muscle tone, short stature, incomplete sexual development, cognitive disabilities, problem behaviors, and a chronic feeling of hunger that can lead to excessive eating and obesity.[7] While there is currently no cure for PWS, early detection allows alleviation and management of most symptoms.

Deficiency, Short and Very-Long Chain Acyl-CoA Dehydrogenase Deficiency, Trifunctional Protein Deficiency. *Hemoglobin disorders*, e.g. sickle cell anemia, Thalassemia, and Hb S/C disease. *Other disorders*, e.g. Congenital hypothyroidism, Biotinidase deficiency, Congenital adrenal hyperplasia, and Cystic fibrosis. See http://www.marchofdimes.com/professionals/14332_1200.asp. See also http://genes-r-us.uthscsa.edu/.

[5] JP Evans, C Skrzynia and W Burke, The complexities of predictive genetic testing, *British Medical Journal*, 2001, **322**: 1052–1056.

[6] ACJW Janssens *et al.*, Predictive genetic testing for type 2 diabetes may raise unrealistic expectations, *British Medical Journal*, 2006, **333**: 509–510.

[7] For more information, see http://www.pwsausa.org/faq.htm (the PWS Association of USA) and the relevant pages at GeneTests.org: http://www.ncbi.nlm.nih.gov/bookshelf/br.fcgi?book=gene&part=pws.

Specific clinical criteria have been defined for the initial diagnosis of PWS from symptoms.[8] The genetic cause of PWS is lack of expression of a group of genes located on the long (q) arm of chromosome 15, between bands 11 and 13. In a process known as imprinting, the maternal copies of five genes in this region are normally silenced by DNA methylation, so that only paternally inherited genes are functional.

Deletion of the15q11-q13 region from the paternal chromosome is the cause of about 70% of PWS cases. Most of the remaining cases of PWS have two maternal copies of chromosome 15 (maternal Uniparental Disomy, UPD) and therefore no active genes in the PWS region. In less than 2% of PWS cases, there are no DNA sequence abnormalities in 15q11-q13. Instead, PWS results from a defect in the imprinting process, resulting in methylation (silencing) of the paternal PWS-region genes.[9]

PWS sufferers are rarely fertile, so almost all PWS cases are *de novo*, but defects in the machinery that performs PWS-region imprinting can be inherited from unaffected parents.[10] For this particular class of individuals, there is an up to 50% risk of PWS in siblings. Thus both chromosomal and methylation tests are often necessary to establish the exact cause of observed PWS-like symptoms. If an imprinting defect is detected, then the methylation patterns of the parents will need to be analyzed in order to verify the possibility of inheritance and assess the risk of PWS in subsequent offspring.

The above example illustrates how genetic testing strategies must be tailored to the biology of each disorder. An additional point to note here is that one of the current tests for PWS checks epigenetic rather than genetic markers. Although the underlying disorder is, by definition, genetic, to date DNA sequencing has not always been the most efficient test option. However, the advent of personal whole-genome sequencing may change this situation.

In the case of PWS, imprinting defects are due to deletions affecting a 4.3 Kbp DNA region known as the PWS Imprinting Center (IC). Recall from Chapter 4 that epigenetic marks are normally erased in the primordial germ cells of embryos. Subsequently, during the production of sperm and egg cells (gametogenesis), a group of genes are differentially

[8] M Gunay-Aygun *et al.*, The changing purpose of Prader–Willi Syndrome clinical diagnostic criteria and proposed revised criteria, *Pediatrics*, 2001, **108**: e92.

[9] J Perk *et al.*, The imprinting mechanism of the Prader–Willi/Angelman regional control center, *EMBO Journal*, 2002, **21**(21): 5807–5814.

[10] K Buiting *et al.*, Detection of aberrant DNA methylation in unique Prader–Willi syndrome patients and its diagnostic implications, *Human Molecular Genetics*, 1994, **3**(6): 893–895.

epigenetically "imprinted" in a sex-specific manner. For these genes, only the maternal or the paternal allele will be expressed in the offspring's somatic cells.[11]

Microdeletions in the PWS IC can be present in a father's maternal chromosome.[12] Because the father's paternal (non-imprinted) chromosome 15 is functional, he would not have PWS himself. But, with a 50% probability, the father may pass his faulty maternal chromosome 15 to an offspring. For the offspring, the faulty chromosome is paternal; its failure to clear the methylation marks in the PWS region results in the lack of PWS-region gene expression and consequent PWS.[13]

From the above description, we see that in principle inherited PWS imprinting center defects can be detected by both DNA methylation tests and diploid sequencing. In practice, tests are initially performed for the far more common cases of large-scale deletions and uniparental disomy. These tests are well-established and widely available. If the results are negative, methlyation testing is available as a specialist service through established labs. In the near future, low-cost and accurate diploid whole-genome sequencing can replace this sequential and effort-intensive process with a single computational assessment of all the PWS-relevant loci.

Very many different genetic testing assays and technologies are currently available. Online resources such as GeneTests.org perform a critical role in informing medical professionals and patients of the latest available tests and their strengths/weaknesses.

In the sections below, we will review a few very common types of genetic tests, but the review is not by any means comprehensive. Rather, our intention is to put the concept and technologies of personal genome sequencing in the context of "traditional" and complementary approaches. In particular, we note that accurate low-cost whole-genome sequencing will be able to substitute for almost all of these tests. A likely scenario is that targeted tests will be used to confirm the results of whole-genome analysis when specific cases are identified.

Biochemical Tests

Tests for "inborn errors of metabolism" are typically performed by measuring the concentration of a metabolite, an enzyme, or an enzyme substrate in either urine or blood samples.

---

[11] See http://www.geneimprint.org/ for articles and links on imprinting.

[12] S Saito *et al.*, Minimal definition of the imprinting center and fixation of a chromosome 15q11-q13 epigenotype by imprinting mutations, *PNAS*, 1996, **93**: 7811–7815.

[13] The details of this process are not yet clear, but see B Kantor *et al.*, Establishing the epigenetic status of the Prader–Willi/Angelman imprinting center in the gametes and embryo, *Human Molecular Genetics*, 2004, **13**(22): 2767–2779.

In recent years, tandem mass spectrometry has become the technology of choice for this purpose because it allows rapid, parallel testing for multiple disorders from a single sample.

Historically, many case-specific biochemical tests have been developed. Many of these tests are both highly sensitive and inexpensive. For example, we noted in our earlier discussion of carrier testing that Cystic Fibrosis (CF) can be difficult to diagnose genetically because of the large number of mutations that can give rise to it. While this obstacle will be removed with the availability of cheap whole-genome sequencing, CF is currently diagnosed more efficiently from sweat chloride levels.

Cytogenetic Detection of Large-Scale Chromosomal Abnormalities

The chromosomes of mitotic cells at metaphase are highly condensed and therefore visible under the microscope. Staining results in distinct chromosomal banding patterns which can be used to identify abnormal numbers of chromosomes, as well as large-scale structural variations (>3 Mbp) such as duplications, deletions, and balanced rearrangements (translocations, inversions). The Giesma stain binds the phosphate groups of A-T rich DNA regions. Staining with Giesma tends to produce dark (G) bands in regions where the chromosome is tightly packaged. In contrast, R (reverse Giesma) banding marks more loosely packed, G-C rich chromosomal regions. C bands mark (mostly centromeric) constitutive heterochromatin.

In fluorescence *in-situ* hybridization (FISH),[14] the presence of specific chromosomes or chromosomal regions is identified through hybridization of fluorescently labeled complementary DNA to denatured chromosomal DNA. Following hybridization, the slide is examined under a fluorescence microscope. Increases or decreases in the number of fluorescing DNA regions mark duplication/deletion events. Fluorescence in unexpected regions may indicate translocation events.

Another approach is to amplify the DNA region of interest using the Polymerase Chain Reaction (PCR). Primers in invariant regions flanking the sequence of interest are used to selectively amplify the target DNA region. Comparison of the length(s) of the resulting PCR product(s) to reference/control samples can identify duplications and deletions. Alternatively, the amplified DNA can be directly sequenced.

---

[14] A variety of FISH assays can reveal different aspects of DNA organization at the single-cell level. Reviewed in EV Volpi1 and JM Bridger, FISH glossary: an overview of the fluorescence *in situ* hybridization technique, *BioTechniques*, 2008, **45**: 385–409.

In Comparative Genomic Hybridization (CGH), the genome of interest is hybridized to a set of reference metaphase chromosomes in competition with control/reference genome. The test and control DNA are labeled with different color fluorescent dyes (e.g. green and red respectively). The hybridization fluorescence ratio indicates the occurrence of insertions or deletions larger than about 3 Mbp within that chromosome. This "traditional" form of CGH is widely used to analyze solid tumors where it is difficult to obtain high quality metaphase chromosomes for FISH.

Instead of whole genomes, Array CGH uses an array of BAC (Bacterial Artificial Chromosome) clones, cDNAs, PCR products, or synthesized oligonucleotides as probes.[15] The relative hybridization at each probe is mapped to the probe's location within the genome, thus providing a much more detailed view than chromosomal CGH. The resolution of Array CGH is determined by the lengths of the probes and their frequency along the genome on the one hand, and cross-hybridization noise on the other. Currently, the resolution is in the order of kilobases. Commercial Array CGH platforms are either targeted to detect specific disease copy number variations, or simply sample the entire genome at roughly equal intervals. Note that CGH and Array CGH cannot directly detect inversions and balanced translocations because these events cause no significant changes in hybridization rates.

Targeted DNA Sequence Analysis

Targeted DNA sequencing typically involves amplification and sequencing of only a specific portion of genomic DNA. Many targeted DNA sequence analysis methods (e.g. single-strand conformation polymorphism and conformation-sensitive gel electrophoresis) avoid the cost of whole genome sequencing at the expense of sensitivity and/or efficiency.

Denaturing high performance liquid chromatography (DHPLC) is a good example of a low-running-cost, high-throughput and sensitive targeted analysis method.[16] DHPLC detects polymorphisms of 1 bp or longer in DNA fragments around 100–1500 bp long. It comprises two steps. First, the DNA segment of interest is amplified using PCR. The amplified

---

[15] See for example C Lee, AJ Iafrate and AR Brothman, Copy number variations and clinical cytogenetic diagnosis of constitutional disorders, *Nature Genetics*, 2007, **39**: S48–S54; AL Beaudet and JW Belmont, Array-based DNA diagnostics: let the revolution begin, *Annual Reviews — Medicine*, 2008, **59**: 113–129.

[16] See TA Sivakumaran, K Kucheria and PJ Oefner, Denaturing high performance liquid chromatography in the molecular diagnosis of genetic disorders, *Current Science*, 2003, **84**(3): 291–296. For a list of HPLC-related papers, see http://insertion.stanford.edu/pub.html.

DNA is then mixed with control (reference) DNA and the mixture is denatured by heat. The separated DNA strands are then allowed to anneal back together at a lower temperature. At locations where the patient DNA varies from the reference sample, heteroduplexes (mis-paired bases) form in the annealed double-stranded DNA fragments. In the second step (chromatography), elution of DNA fragments is detected by UV absorbance. The presence of one or more heteroduplexes produces additional peaks in the elution profile (compared to the reference).

DHPLC indicates only the presence of mutations/polymorphisms in DNA. It does not specify what they are. If necessary, additional sequencing has to be performed to establish the nature of the variation(s).

In addition to detection of sequence polymorphisms, DHPLC can be adapted to detect DNA methylation. Treatment of DNA with sodium bisulfite converts cytosine residues to uracil, but leaves methylated cytosine residues unaffected. Uracil is substituted by thymidine during PCR amplification. However, methylcytosine appears as cytosine after PCR. The DNA methylation-specific sequence changes introduced by bisulfite treatment can be characterized by comparison with untreated DNA.

Direct sequencing of bisulfite-treated DNA would be most informative, but to date the need for relatively large amounts of DNA in sequencing has limited its applicability. This is likely to change with the advent of new sequencing technologies (reviewed below). But for the moment bisulfite treatment followed by DHPLC offers a more efficient means of detecting DNA methylation in well-defined genomic regions.[17]

Recall from Chapter 3 that — because of the way recombination shuffles DNA segments — small-scale sequence variations in nearby genomic locations tend to co-occur within populations. Thus, for a given population, one can define "haplotypes", i.e. DNA sequence regions within which common sequence variations are highly correlated. Single Nucleotide Polymorphisms (SNPs) can provide excellent "tags" representative of haplotype regions.

Tag SNP array chips provide genome-wide information at a fraction of whole-genome sequencing. However, it is important to keep in mind that haplotypes — and therefore tag SNPs — can only capture commonly occurring sequence variations. Rare alleles are, by definition, difficult to capture or correlate to other events. For this reason, a number of current SNP-array products include additional custom probes that can target known rare alleles of interest.

---

[17] For example, DHPLC has been successfully demonstrated to detect methylation of tumor suppressor genes, see B Betz *et al.*, Denaturing High-Performance Liquid Chromatography (DHPLC) as a reliable high-throughput prescreening method for aberrant promoter methylation in cancer, *Human Mutation*, 2004, **23**: 612–620.

For common alleles occurring in >5% of the population, studies suggest that in the region of 1 million tag SNPs will capture nearly every common variant.[18] A number of commercial products already provide SNP arrays at and beyond this order of complexity.

A powerful application of SNP arrays is in pharmacogenetics, i.e. the prediction of individual responses to drugs. For example, the anti-tuberculosis drug isoniazid causes liver damage in patients with a variety of SNPs in their *NAT2* gene.[19] Ideally, isoniazid dosage should be adjusted within a three-fold range depending on the patient's *NAT2* genotype.[20]

Until whole genome sequencing becomes sufficiently inexpensive, SNP array chips can be used to identify such drug sensitivity and adverse reaction alleles. Once established, such information could be added to the individual's health records and used throughout his or her life.

De novo Whole-Genome DNA Sequencing

It may soon be more cost-effective to sequence an individual's entire genome instead of performing a series of targeted analyses. Whereas multiple targeted tests may be required throughout the life of an individual, personal whole-genome sequencing need only be performed once. Moreover, personal genomic sequencing may identify relevant rare variants not included in standardized tests. Viewed at the pathway, process and system levels, some rare variants may turn out to cluster in disease-associated pathways. Such rare sequence variants may either predispose or protect against disease. Their discovery will help both the sequenced individual and also general understanding of the mechanism(s) of the disease.

As we saw in the introduction, the cost of genome sequencing has been dropping exponentially since its advent. Indeed, during the course of the international human genome project, the cost of DNA sequencing dropped 100-fold from $10 to $0.1 per finished base.[21]

[18] JC Barrett and LR Cardon, Evaluating coverage of genome-wide association studies, *Nature Genetics*, 2006, **38**(6): 659–662.

[19] K Fukino *et al.*, Effects of N-acetyltransferase 2 (NAT2), CYP2E1 and Glutathione-S-transferase (GST) genotypes on the serum concentrations of isoniazid and metabolites in tuberculosis patient, *Journal of Toxicological Sciences*, 2008, **33**(2): 187–195.

[20] M Kinzig-Schippers *et al.*, Should we use N-Acetyltransferase Type 2 genotyping to personalize isoniazid doses? *Antimicrobial Agents and Chemotherapy*, 2005, **49**(5): 1733–1738.

[21] JA Schloss, How to get genomes at one ten-thousandth the cost, *Nature Biotechnology*, 2008, **26**(10): 1113–1115; J Shendure *et al.*, Advanced sequencing technologies: methods and goals, *Nature Reviews Genetics*, 2004, **5**(5): 335–344.

The total cost of the international human genome project is estimated at about $3 billion.[22] In contrast, the cost of sequencing an individual human genome using "next-generation" parallel sequencing dropped to less than $500,000 in 2008[23] and to less than $50,000 in 2009.[24] Current NHGRI funding is aimed at delivering $1000 personal genomes by 2012.[25]

There are several reasons for the dramatically lower costs of the emerging generation of sequencing technologies. We will review some of these in the next section. A key enabling factor is that the human genome has already been sequenced, both as an aggregate haploid sequence from multiple individuals[26] (the public and commercial human genome projects) and also in high-quality diploid form for a single individual.[27]

The availability of these *de novo* assembled genomes greatly simplifies the task of sequencing further individuals. The reason is that, so far, it has been difficult to sequence DNA fragments much larger than about 1 or 2 Kbp (discussed below). Since the human genome includes many repeat sequences of comparable or longer length (see Chapter 3), it is often difficult to determine the order and overlap pattern of sequenced fragments. The commercial[28] and public[29] human genome projects investigated radically different approaches to addressing this issue.

Below, we provide a brief outline of the manner in which the three reference human genomes were sequenced. Our aim is not to review the history of the *de novo* human genome projects,[30] but rather to highlight their role in facilitating personal genomics and to provide a point of reference for our discussion of the emerging technologies in the next section.

---

[22] FS Collins, M Morgan and A Patrinos, The human genome project: lessons from large-scale biology, *Science*, 2003, **300**(5617): 286–290.

[23] J Wang *et al.*, The diploid genome sequence of an Asian individual, *Nature*, 2008, **456**(7218): 60–65.

[24] M Allison, Illumina's cut-price genome scan, *Nature Biotechnology*, 2009, **27**(8): 685.

[25] A Coombs, The sequencing shakeup, *Nature Biotechnology*, 2008, **26**(10): 1109–1112.

[26] International Human Genome Sequencing Consortium, Finishing the euchromatic sequence of the human genome, *Nature*, 2004, **431**: 931–945; S Istrail *et al.*, Whole-genome shotgun assembly and comparison of human genome assemblies, *Proceedings of the National Academy of Sciences of USA*, 2004, **101**(7): 1916–1921.

[27] Craig Venter's genome, see http://www.jcvi.org/cms/research/projects/huref/overview/ and links therein.

[28] JC Venter *et al.*, The sequence of the human genome, *Science*, 2001, **291**: 1304–1351.

[29] The International Human Genome Sequencing Consortium, Initial sequencing and analysis of the human genome, *Nature*, 2001, **409**: 860–921.

[30] For a brief history of the human genome project, see http://tinyurl.com/HGP-history. For a brief review of the history of DNA sequencing, see CA Hutchison, DNA sequencing: bench to bedside and beyond, *Nucleic Acids Research*, 2007, **35**(18): 6227–6237.

The early years of the international (public) human genome project were spent developing genetic[31] and physical[32] genomic maps that provided marker sets. As noted above, current technologies cannot sequence entire chromosomes as one continuous piece. The markers established by the international human genome project were essential in allowing the ordering of sequence fragments along chromosomes.

To sequence the human genome, DNA from anonymous volunteers was first fragmented into DNA segments of about 70 Kbp to 230 Kbp each and inserted into Bacterial Artificial Chromosomes (BACs)[33] for amplification. Known genomic markers (and to a lesser extent sequenced clone ends) were then used to tile each chromosome with a series of overlapping BAC clones.

The availability of ordered, overlapping BAC clones simplified the problem of assembling the full genome from small (<1 Kbp) sequence reads to the problem of assembling the sequences within each (~150 Kbp) BAC clone. This approach has become known as hierarchical or clone-based assembly.

Figure 6.1 summarizes the hierarchical assembly approach. The tick marks along the "mapped clones" represent genomic markers. The matching end-overlaps of the clones are highlighted. The assembly of sequence fragments (from one BAC clone) into a contiguous sequence (a "contig") is schematically shown at the bottom. Note that for simplicity only a few sequence fragments are shown.

The number of sequence fragments covering a given portion of the genome is referred to as "fold-coverage" or "redundancy". In the schematic example in figure 6.1, the fold-coverage in the central region is three-fold in some areas and two-fold in others. In practice, outside of some difficult-to-sequence (usually repeat-rich) portions of the genome, the human genome project typically achieved a coverage of 8–10 fold.

The current publicly available aggregate (i.e. multi-individual) reference sequence is a composite of the public and commercial assembled genomes.[34] As we will see shortly, the commercial and public projects used the same core sequencing technology, but their initial approaches to sequence assembly were sharply different.

---

[31] JC Murray *et al.*, A comprehensive human linkage map with centimorgan density, *Science*, 1994, **265**(5181): 2049–2054; H Donis-Keller *et al.*, A genetic linkage map of the human genome, *Cell*, 1987, **51**(2): 319–337.

[32] TJ Hudson *et al.*, An STS-based map of the human genome, *Science*, 1995, **270**(5244): 1945–1954.

[33] Initial analysis was performed using clones ranging from about 40 kilobases (cosmids) to several hundred kilobases (yeast artificial chromosomes). In the late 1990s, it was decided to standardize the clones for the bulk of the genome sequence using ~150 Kbp BAC libraries prepared from two anonymous DNA donors.

[34] See http://www.ncbi.nlm.nih.gov/mapview/static/humansearch.html for details.

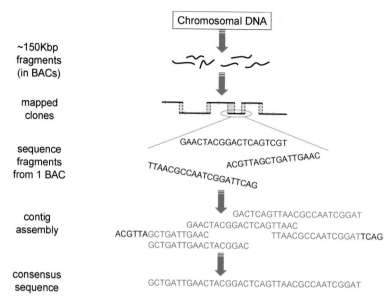

Figure 6.1:   Hierarchical sequencing and assembly.

The Celera (commercial) human genome project, which pooled the DNA of five individuals, adopted a strategy known as whole-genome shotgun assembly (WGSA). WGSA skips the time-consuming process of identifying genomic markers, developing a physical map of the genome, and establishing a library of BAC clones that tile each chromosome. Instead, Celera fragmented genomic DNA into 2, 10 and 50 Kbp pieces, which were integrated into a plasmid library. Each plasmid insert was sequenced at both ends.

The sequences at the opposite ends of a given plasmid insert are called mate pairs or paired ends. Because the approximate distance between mate pairs is known a priori, and the orientation of mate pairs with respect to each other is fixed, mate-pair sequences are much easier to align.[35] The use of paired-end sequences in shotgun sequencing is sometimes referred to as double-barreled shotgun sequencing.

During the assembly processes, only paired ends that fall within three standard deviations of the average plasmid insert size are accepted as true alignments. Use of mate pairs at three different length scales (2, 10, and 50 Kbp in the case of the Celera human genome), allows

---

[35] JC Roach *et al.*, Pairwise end sequencing: a unified approach to genomic mapping and sequencing, *Genomics*, 1995, **26**: 345–353.

assembly algorithms to overcome misalignments due to repeat sequences of various lengths. Nonetheless, the Celera assembly had more difficulty with repeat sequences than the public assembly.[26] On the other hand, paired-end WGSA sequencing avoided some orientation and location errors in the public sequence[26] and reduced the need for a physical map.

The assembly algorithms developed for whole-genome shotgun sequencing use multiple heuristic measures to fill in gaps and resolve ambiguities in "reasonable" computing time.[36] Nonetheless, each assembly-run of the WGSA human genome took about 20,000 CPU hours on a dedicated processor cluster.[28] *De novo* sequencing with next-generation sequencing technologies (see next section) currently requires at least an order of magnitude more processing time.[37] However, for personal genomes, sequences can be aligned to the existing reference genomes at much lower costs.

The technology used to sequence the human genome in both of the above projects is known as Sanger or chain termination sequencing. In this type of sequencing, the DNA fragment to be sequenced (the template) is replicated using DNA Polymerase and a primer that starts the replication process at one end of the template. The process is similar to DNA replication during the cell cycle. A key difference is the use of dideoxynucleotide triphosphates (ddNTPs) in addition to the normal nucleotides (dNTPs) found in cells.

ddNTPs are identical to natural dNTPs except that they contain a hydrogen group on the 3' carbon instead of a hydroxyl (OH) group. When incorporated into a sequence, ddNTPs prevent the addition of further nucleotides to the growing DNA copy. This is because, in the absence of a hydroxyl group, a phosphodiester bond cannot form between the ddNTP and the next incoming nucleotide. The upshot is that the replication process is terminated at the point where a ddNTP is incorporated into the replicate DNA.

In current implementations of chain termination sequencing, the four ddNTPs (corresponding to A, C, G, T nucleotides) are labeled with different color fluorescent dyes. The sequencing process is performed with millions of copies of the template DNA. As each template is copied, a ddNTP is randomly incorporated into the replicating chain at one of the occurrences of the corresponding nucleotide.

Figure 6.2 illustrates the outcome. The sequence at the top represents the initial portion of an example template (primer binding sequence not shown; 3' end at left, 5' at right).

---

[36] See for example X Huang *et al.*, PCAP: a whole-genome assembly program, *Genome Research*, 2003, **13**: 2164–2170.

[37] M Pop and SL Salzberg, Bioinformatics challenges of new sequencing technology, *Trends in Genetics*, 2008, **24**(3): 142–149.

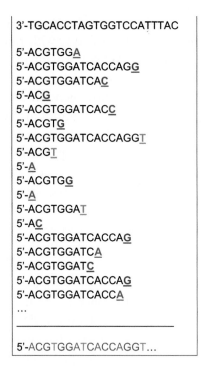

Figure 6.2:   How chain termination signals reveal the DNA sequence.

The sequences below the template show example replicates that have terminated upon incorporation of a fluorescent ddNTP (boldface underlined letters). The line at the bottom shows how the consensus alignment of the synthesized fragments reveals a sequence of ddNTPs complementary to the template DNA sequence. Each sequenced piece of DNA is referred to as a sequencing read (a "read" for short).

In current implementations of Sanger sequencing, the positions of the ddNTPS in the synthesized fragments are identified by capillary gel electrophoresis. DNA is negatively charged, so that it will migrate towards the positive pole in an electric field. Smaller DNA fragments migrate faster than longer fragments. Optimized capillary gels can distinguish between DNA fragments differing in size by just one nucleotide. Thus the order in which ddNTPS reach a fluorescence laser near the positive terminal of the gel defines their location along the template.

Separation by electrophoresis puts a limit on the maximum length of DNA fragments that can be sequenced with capillary electrophoresis. A DNA fragment 2 bp long is twice the

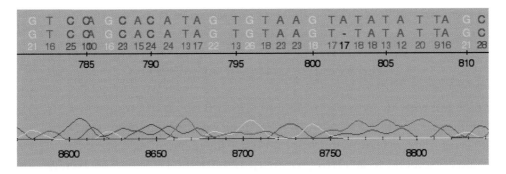

Figure 6.3: An example DNA sequencing chromatogram.

mass of a single nucleotide, so the two can be distinguished easily as they migrate towards the positive electrode. But DNA fragments 1000 bp and 1001 bp differ in weight by only 0.1%, making their separation much more difficult. As a result, DNA sequences recovered by Sanger sequencing are typically 500 to 850 bp long, although they can extend to ~1250 bp.

In addition to poor separation between successive fragments, detection of the fluorescent signal can be noisy due to other effects such as overlapping color wavelengths of the four ddNTPs. Base-pair calling algorithms use a variety of quality measures to translate fluorescence chromatographs into nucleotide sequences. Figure 6.3 shows an example chromatogram and associated base-pair calls.[38]

The overlapping curves at the bottom are the fluorescence recordings for each nucleotide. Each peak represents normalized fluorescence measured at one nucleotide position. The taller a peak, the stronger the fluorescence signal at that point. But the ratio of signal levels (with respect to the other signal levels at a given location, and also compared to the reading for the same base in neighboring locations) matters more than absolute values.

The upper row of letters indicates the reference sequence. The line below it shows the bases called in the fragment being sequenced. The numbers below each letter are quality scores calculated by a program called Phred.[39] Phred uses a number of empirically determined criteria (e.g. peak height and spacing, and amplitude ratio to next highest signal) to

---

[38] Viewed in the open-source Hawkeye assembly viewer: http://amos.sourceforge.net/hawkeye.

[39] B Ewing and P Green, Base-calling of automated sequencer traces using Phred. II. Error probabilities, *Genome Research*, 1998, **8**(3): 186–194.

assign an error probability to each call. The reported score is log transformed according to the formula:

$$PherdScore = -10 \log_{10} (errorprobability).$$

Thus, a base call with an error probability of 1 in 100 (i.e. 0.01) results in a Phred score of 20. A Phred score of 20 is typically taken as "high quality". Ignoring systematic, assembly, and other errors, at a given nucleotide location five independent calls each with a Phred score of 20 would correspond to an error probability of 0.01^5 or 10^{-10}; i.e. on average one would expect less than one incorrectly called nucleotide per diploid human genome.[40]

If we assume three-fold coverage instead of five-fold, the above overall error probability would be one in every million bases, i.e. thousands of erroneous base-pair calls in a single human genome. This consideration highlights the trade-off between base-call quality and fold-coverage. One aspect of the trade-off is that sequencing costs can be balanced against the accuracy/reliability of delivered sequence. Another is that sequencing technologies with lower base-calling accuracy can be competitive if they are sufficiently low-cost to allow a compensatory high level of redundancy.

In 2007, the individual genomes of Craig Venter[41] and James Watson[42] were sequenced using Sanger and next-generation (Roche 454) technologies respectively. Venter's genome, which was sequenced in fragments of 500–1000 bp in length, was assembled *de novo* (i.e. without reference to the existing human genome sequence) by WGSA. Watson's genome, which was sequenced in reads averaging ~250 bp, was largely assembled by alignment of reads to the existing human genome. Both genomes were sequenced to ~7.5-fold coverage. This allowed detection of ~75% of heterozygote alleles in both cases. As discussed later in this chapter, 13-fold coverage would be needed to identify 99% of all heterozygote alleles.

Whereas Venter's *de novo* genome assembly revealed extensive multi-nucleotide variations (insertions, deletions, translocations, duplications, etc.) at scales ranging from 2 bp to tens of kilobases, Watson's assembly could only detect multi-nucleotide variations up to the read length (~250 bp). This observation highlights the need for paired-end sequencing with multiple gap lengths (not used for Watson's genome). Detection of large-scale (structural)

---

[40] In practice, the public human genome — which was sequenced at ~8 to 10-fold coverage — achieved an overall accuracy of about 99.99% (i.e. an error rate of 1 in 10,000 bases).
[41] S Levy *et al.*, The diploid genome sequence of an individual human, *PLoS Biology*, 2007, **5(10)**: e254.
[42] DA Wheeler *et al.*, The complete genome of an individual by massively parallel DNA sequencing, *Nature*, 2008, **452**(7189): 872–876.

variations requires additional comparative genomic hybridization (CGH) microarray analysis, as was performed for both the Venter and Watson genomes.

The above comparisons between the Sanger-sequenced Venter genome and the short-read sequenced Watson genome should be viewed in the context of the time and cost of sequencing each genome.[43] Venter's genome sequence cost in the region of $100 million and took four years to complete. Watson's genome was sequenced in just a few months for about one hundredth of the cost. Both Watson and Venter's sequencing runs were completed in 2007. In 2009, Complete Genomics is promising to sequence entire human genomes using short, paired-end reads in a matter of days, and for as little as $5000 per genome.[44] Thus, the use of next-generation technologies for personal genome re-sequencing is becoming increasingly compelling.

Next-Generation (Parallel) Re-sequencing Technologies

New DNA sequencing technologies are evolving rapidly and in diverse ways.[45] Our review here is therefore not concerned with the rapidly changing technical details. Instead, we review the common features of emerging ("next-generation") parallel sequencing technologies. We will discuss why and how sequencing an entire human genome in less than a week and for less than $1000 is likely to become reality within the next five years. We will also consider possible limitations of emerging technologies, and their implications for personalized medicine.

The US National Human Genome Research Institute (NHGRI) is currently funding some 50 next-generation sequencing technology projects at a cost of about $100 million.[21] The stated aim of the initiative is to deliver $1000 human genomes by 2014. The Archon X-prize for genomics (http://genomics.xprize.org/) is currently offering a $10 million award to any team that:

> "can build a device and use it to sequence 100 human genomes within 10 days or less, with an accuracy of no more than one error in every 100,000 bases sequenced, with sequences accurately covering at least 98% of the genome, and at a recurring cost of no more than $10,000 per genome."

---

[43] M Wadman, James Watson's genome sequenced at high speed, *Nature*, 2008, **452**(7189): 788.

[44] http://www.completegenomicsinc.com/pages/materials/CompleteGenomicsLaunchPressReleasel.pdf

[45] In late 2008, five companies were offering commercial next-generation technologies: Applied Biosystems (SOLiD technology), Roche (454 FLX), Illumina (Genome Analyzer), Dover Systems (Polonator G.007), and Helicos (HeliScope). At least six more companies are actively developing next-generation technologies: Pacific Biosciences, Visigen Biotechnologies, LaserGen, Inc., Intelligent Bio-Systems, Complete Genomics, and Oxford Nanopore Technology. See Ref. 25 for details.

As of July 2009, seven teams have registered to compete for the prize, but no one has yet set a date for an attempt.

Automated, high-throughput chain termination (Sanger) sequencing is currently performed in up to 384 wells in parallel to deliver ~86 Kbp/hour at a cost of about $0.5 per kilobase. Thus, to sequence an entire diploid human genome today using Sanger sequencing would cost in the region of $500K and would take more than eight years on a single sequencing machine.

One way to speed up Sanger sequencing is to parallelize capillary electrophoresis beyond 384. A 384-microcapillary system on glass wafers was successfully demonstrated in 2002.[46] More recently, an integrated Sanger sequencing system using microfluidic flow cells and microcapillaries etched into 100 mm glass wafers was demonstrated.[47]

Such microsystems use well-established manufacturing methods developed by the semiconductor industry and offer two potential advantages. First, they can accommodate large numbers of microcapillaries to enable further parallelization. Second, microsystems use hundreds of times less reagents and template DNA, therefore allowing very low-cost sequencing.

Thus, in principle, it may be possible to use an array of miniature Sanger sequencers to achieve gains of a few orders of magnitude throughput in Sanger sequencing. However, the technology is still embryonic and not expected to become commercially viable in the next few years. Partly because of the challenges of parallelizing Sanger sequencing, most current research and development is focused on technologies that do not require electrophoresis.

The new sequencing technologies all have one feature in common. Instead of performing a few hundred or a few thousand reads at a time (as in capillary and microcapillary Sanger sequencing), millions of reads may be performed in parallel. This massive parallelism is achieved through micro-compartmentalization.

The techniques used to generate and populate the micro-compartments vary considerably among projects, but the underlying concept is the same. In the current and near-term technologies, single-stranded DNA fragments are first amplified within individual micro-compartments using PCR. Next, sequencing reactions are performed on these pools of identical DNA fragments, resulting in good signal-to-noise ratios.

[46] CA Emrich *et al.*, Microfabricated 384-lane capillary array electrophoresis bioanalyzer for ultrahigh-throughput genetic analysis, *Analytical Chemistry*, 2002, **4**(19): 5076–5083.
[47] RG Blazej, P Kumaresan and RA Mathies, Microfabricated bioprocessor for integrated nanoliter-scale Sanger DNA sequencing, *Proceedings of the National Academy of Sciences of USA*, 2006, **103**(19): 7240–7245.

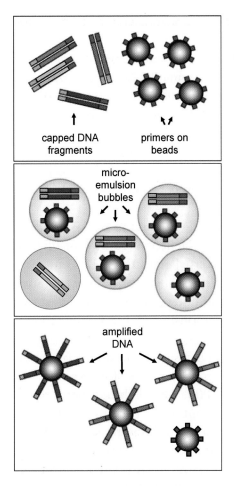

Figure 6.4: Emulsion PCR.

Figure 6.4[48] illustrates a widely used approach to compartmentalized DNA amplification known as emulsion PCR.[49] Here, genomic DNA is first sheared into fragments and capped at both ends with adaptor DNA sequences. The capped DNA fragments, primers and dNTPs are added to a mixture of oil and water (top panel). One of the two primers is

---

[48] Figure based on J Shendure and H Ji, Next-generation DNA sequencing, *Nature Biotechnology*, 2008, **26**(10): 1135–1145.

[49] D Dressman *et al.*, Transforming single DNA molecules into fluorescent magnetic particles for detection and enumeration of genetic variations, *Proceedings of the National Academy of Sciences of USA*, 2003, **100**(15): 8817–8822.

attached to 1 μm-sized beads, so that future PCR reactions will result in bead-attached amplicons.

Vigorous stirring of the mixture results in the formation of a water-in-oil micro-emulsion (i.e. "bubbles" of aqueous solution in oil). Low DNA and bead concentrations are chosen so that the probability of having more than one template per compartment is negligibly low (middle panel). PCR reactions take place in compartments where template and primer are both present. Following amplification, the magnetically charged beads carrying the amplified DNA can be isolated and used for sequencing (bottom panel).

Each cluster of amplified DNA is sequenced individually in parallel with all the other clusters. Typically, the DNA clusters are immobilized on a surface to facilitate successive reads.

At least two emerging technologies aim to bypass template amplification, and sequence single DNA molecules directly. Helicos uses a form of Fluorescence Resonance Energy Transfer (FRET) and total internal reflection (TIR) microscopy to enable measurement of fluorescence signals from single nucleotide incorporation events in arrays of single-stranded DNA templates randomly tethered to the surface of a quartz slide.[50] The technology was successfully demonstrated in 2008 by sequencing the 6.4 Kbp of the M13 viral genome.[51]

In the summer of 2009, Helicos sequenced 90% of the genome of its founder Stephen Quake using a single machine over a period of four weeks.[52] The total cost of sequencing was estimated to be less than $50,000. Although individual reads are only 24–70 bp long and have a raw total error rate of about 3.5% (roughly one base pair per read), the low cost and high throughput of the technology was used to achieve 28-fold average coverage.

Nanopore sequencing provides single-molecule detection by electrophoretically driving individual DNA molecules (in solution) through a nanometer-scale pore. Both protein and synthetic pores are being investigated. Nanopores are sufficiently small that DNA molecules can only move through them single-file, one nucleotide at a time. Each nucleotide within a nanopore can be detected by a variety of means, such as changes in conductance.

A particularly attractive feature of nanopore sequencing is that read lengths can be kilobases long. Another potentially attractive feature is that nanopore sequencing does

---

[50] I Braslavsky *et al.*, Sequence information can be obtained from single DNA molecules, *Proceedings of the National Academy of Sciences of USA*, 2003, **100**: 3960–3964.

[51] TD Harris, Single-molecule DNA sequencing of a viral genome, *Science*, 2008, **320**: 106–109.

[52] D Pushkarev, NF Neff and SR Quake, Single-molecule sequencing of an individual human genome, *Nature Biotechnology*, 2009, **27**: 847–850.

not require reagents and so could potentially enable very low-cost sequencing. Moreover, sample preparation is minimal, sequencing is performed on single DNA molecules, and the potential for parallelism and high throughput is great. However, nanopore technology is still in its infancy, and multiple challenges in manufacture, control and measurement remain.[53]

The sequencing chemistry used by each technology provider is different. The ABi SOLiD system and Dover Systems' Polonator use "sequencing by ligation".[54] As the name implies, this approach identifies template sequences by competitive binding of specially designed fluorescently labeled oligonucleotides using DNA ligase. In the SOLiD system, each ligation event detects two nucleotides at a time. Overlapping successive ligation events improve accuracy per base calls, but read lengths have so far been limited to ~50 bp. The Polonator uses single nucleotide ligations from the two ends of 13 bp DNA fragments.

Roche, Illumina and Helicos all use DNA polymerase to perform "sequencing by synthesis". However, each company uses a different chemistry. Illumina uses a propriety reversible chain terminating process, which allows multiple nucleotides to be read sequentially from the same DNA template (one nucleotide per step).

Helicos uses non-terminating fluorescently labeled dNTPs. After an incorporated base has been imaged, the fluorescent tag is removed chemically and the cycle repeated. Difficulties associated with cleaving fluorescent tags have so far limited the read lengths for both Illumina and Helicos. Roche uses pyrosequencing,[55] in which modified nucleotides release a pyrophostate upon incorporation into the replicating sequence. This approach has allowed them average read lengths of up to 500 bp so far.

Since the Helicos and Roche approaches do not use chain-terminating nucleotides, they tend to incorporate multiple nucleotides at homopolymers (base repeats such as AAA, CC, or GGGG). The number of incorporated bases has to be estimated from the fluorescence signals, causing potential errors. In contrast, the primary source of base-call errors in the Illumina technology is substitution.

All of the technologies discussed above are evolving rapidly and will likely overcome any remaining challenges in the near future. The important point for our consideration here is

---

[53] D Branton *et al.*, The potential and challenges of nanopore sequencing, *Nature Biotechnology*, 2008, **26**(10): 1146–1153.

[54] J Shendure *et al.*, Accurate multiplex polony sequencing of an evolved bacterial genome, *Science*, 2005, **309**: 1728–1732.

[55] M Ronaghi, M Uhlén and P Nyrén, A sequencing method based on real-time pyrophosphate detection, *Science*, 1998, **281**: 363–365; see also P Nyrén, The history of pyrosequencing, *Methods in Molecular Biology*, 2007, **373**: 1–14.

that in evaluating the rates and types of possible sequencing errors in personal genomes, the characteristics of the sequencing technology used must be taken into account.

Sequencing Challenges for Personal Genomics

A shared characteristic of the new generation of sequencing technologies has so far been shorter read lengths than Sanger sequencing. Shorter read lengths make the assembly of repeat-rich sequences difficult. Moreover, more reads are required to achieve a desired fold-coverage. However, read lengths on commercially available platforms have been increasing steadily. For example, the latest SOLiD sequencers from ABi (SOLiD 3) deliver read lengths of 2×50 bp for paired-ends[56] compared to 35 bp in the previous version of the same system. The longest non-Sanger reads are currently offered by Roche's 454 sequencer, which currently has an average read length of 400–500 bp, up from 250 bp in the previous year.[57]

Increases in read lengths will also help in the genetic characterization of the "microbiome" of the human body. As we noted in earlier chapters, the human microbiome plays an essential role in human physiology, is extremely diverse, and varies greatly from individual to individual. For these reasons, the personal microbiome is likely to play a key role in personalized medicine. Sequencing can be used to enumerate the types and numbers of the many microbial species in organs such as the gut and the skin. But because multiple bacterial species are inextricably mixed together in the body, the mixture of genomes is usually shotgun sequenced as one sample. Assembly algorithms are then used to assign contigs to different species.[37] Shorter reads generally produce shorter contigs, which in turn make it more difficult to uniquely assign contigs to species. However, read lengths of 200 bp or longer appear to generate good results, so at least some of the new sequencing technologies will be well placed for this application.

At present, the raw per base calling accuracy of (next-generation) parallel sequencing is about ten-fold less than Sanger sequencing;[59] as a result correspondingly higher fold-coverage is needed to generate sufficiently accurate finished base calls. Recall from our discussion in the "*de novo* sequencing" section that a typical cut-off frequency for a good read is a Phred score of 20, corresponding to an error probability of 1%. Ignoring systematic, assembly, and other errors, five calls for the same base at this error rate will give an error probability of the order of one base per diploid human genome (in practice error rates are much higher).

---

[56] October 1, 2008 press release by Applied BioSystems, see http://press.appliedbiosystems.com.

[57] October 1, 2008 press release at http://www.roche.com/media/media_releases/med_dia_2008-10-01.htm.

Lower redundancy dramatically increases the expected number of errors per genome. High redundancy sequencing increases both the cost of sequencing and the total sequencing time.

For medical sequencing, high base-calling accuracy is needed for both alleles so that homo and heterozygous alleles can be distinguished. Figure 6.5 shows the fraction of the diploid human genome that would be covered at both alleles by one to five 31 bp reads as a function of average sequencing redundancy.[58] To cover 99% of the diploid genome with five reads requires at least 25-fold coverage.

The curves for figure 6.5 were generated without taking DNA repeat patterns and chromosomal structure into account. In practice, about 2% of the human genome is extremely difficult to sequence accurately and repeat-rich sequences will require higher fold-coverage. Aneuploid genomes (e.g. in cancer cells) will also require higher sequencing redundancy.[58] Thus, in the short term, it may be necessary to re-sequence any medically important DNA sequence variations found in personal genomes.

A potentially important consideration in the use of second-generation sequencing technologies is that the different sequencing chemistries used by the various technologies lead to

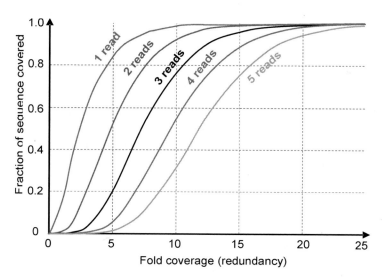

Figure 6.5: The relationship between average coverage and the fraction of sequence covered by 1–5 reads.

[58] Figure reproduced from MC Wendl and RK Wilson, Aspects of coverage in medical DNA sequencing, *BMC Bioinformatics*, 2008, **9**: 239.

differences in the types of mis-reads that can happen.[59] As we saw in the previous section, for the technologies developed by Roche and Helicos, estimating the length of a homopolymer sequence can be noisy, resulting in short insertion–deletion errors. For the SOLiD, Polonator and Illumina technologies, substitutions are the most common error. To be useful for medical applications, reliable models of error probability at the single-base and assembly levels will be needed for each technology (cf. Phred scores for Sanger sequencing base calls).

Recall from Chapter 3 that multi-base DNA sequence variations are frequent and range in size from 2 bp to megabases. Such variations include insertions, deletions, inversions and translocations. The largest scale sequence variations currently need to be identified by methods such as array comparative genomic hybridization (aCGH). For sequence variations shorter than those detected by aCGH and longer than read lengths, (say a few tens of bases to a few tens of kilobases), it is necessary to perform paired-end sequencing.

An insertion or deletion is hypothesized when uniquely aligned mate pairs are found to be an unexpected distance apart (typically more than three standard deviations of the expected mate-pair gap). Inversions are detected when the orientations of multiple overlapping mate pairs are discordant.

Shorter reads result in fewer uniquely aligned end sequences. While nearly 99% of Sanger reads can usually be aligned uniquely, typically only about half of short-read mate pairs are aligned with high confidence. Partly for this reason, predicted sequence variations are usually based on supporting indications from multiple mate pairs. Moreover, mate pairs at several different gap lengths are required in order to span repetitive sequences of varying lengths. The need for highly redundant paired-end reads increases costs and reduces throughput. However, multi-base and structural variations play a key role in genomic diseases (especially cancer), so the ability to characterize them is particularly valuable in personal genomics. All the current next-generation parallel sequencing technologies allow paired-end sequencing with multiple gap lengths.

---

[59] Reviewed in J Shendure and H Ji, Next-generation DNA sequencing, *Nature Biotechnology*, 2008, **26**(10): 1135–1145; also, RA Hold and SJM Jones, The new paradigm of flow cell sequencing, *Genome Research*, 2008, **18**: 839–846.

CHAPTER 7

Measuring the Impact of Environmental Factors on Health

We saw in Chapter 3 that genetic differences among individuals lead to differences in cellular biochemistry, which in turn endow each individual with a personal disease-susceptibility profile. Environmental and life-history factors act on an individual's personal susceptibility profile and determine whether a disorder occurs, the time of onset, the severity of the disease, and response to treatments (see discussions in Chapters 4 and 5). In Chapter 6, we observed that the emerging generation of DNA sequencing machines will soon allow characterization of personal susceptibility profiles. How can we go from susceptibility to diagnosis, prevention and treatment?

This chapter reviews the technologies used to quantify the health status of individuals. The measured quantities — called biomarkers — range from organ physiology and anatomical structure to RNA/protein/metabolite abundance. While in the past biomarker technologies have tended to focus narrowly on specialist diagnosis and staging of life-threatening disorders, there is a trend towards non-invasive, low-cost and low-impact means of monitoring a broad range of physiological and cellular processes.

The sections below start with some definitions and clarifications of key concepts. We will then review established and emerging biomarker technologies, and end with a discussion of the challenges ahead. We will discuss how these measures can be used to infer an individual's health status and devise a personal healthcare strategy in the next two chapters.

Selecting Biomarkers

A biomarker is defined as "a characteristic that is objectively measured and evaluated as an indicator of normal biologic processes, pathogenic processes, or pharmacologic responses to a therapeutic intervention".[1] Sometimes, when a direct measure of a process of interest is not feasible, a surrogate biomarker is used instead.

A surrogate marker is simply "a biomarker intended to substitute for a clinical endpoint".[1] For example, tumor size might be used as a surrogate predictor of mortality in cancer patients. Surrogate markers are particularly useful when clinical end points are undesirable during trials (e.g. relapse, mortality), or difficult to measure directly (e.g. early symptoms of slow-developing diseases such as Alzheimer's).

At present, the relationship between many surrogate markers and associated diseases is poorly understood. This has led to questions about the value of surrogate markers.[2] Is blood cholesterol level a good predictor of cardiovascular disease and stroke? Can the performance of a drug be measured using surrogate markers alone? These questions apply to the future use of biomarkers in personalized medicine as much as they apply to current practices, but with one important difference, as follows.

As systems biology approaches characterize the networks of molecular interactions that determine cellular and organ function, the relationship between the available biomarkers and health outcomes becomes better understood. Moreover, characterized network components and interactions downstream of a dysfunctional gene product will provide new candidate surrogate markers whose relationship to the disorder is mechanistically understood. Thus, observational statistics will increasingly be replaced by mechanistic and causal models. As a result, the predictive power of (surrogate) biomarkers will be better characterized, and the conditions and contexts under which they are good or poor predictors will be enumerated.

In the remainder of this chapter, we will assume that the biomarkers that will be used in future to predict and track disorders are selected on the basis of their known roles in the networks of molecular interaction perturbed by the disorder. Given this assumption, we can set aside concerns about surrogate biomarkers and focus on a discussion of biomarkers in the context of personalized medicine.

Finally, although biomarkers are used widely in molecular epidemiology and environmental biomonitoring, these applications of biomarkers are beyond the scope of our discussions and not discussed in this chapter.[3]

---

[1] AJ Atkinson *et al.* (the NIH Biomarkers Definitions Working Group), Biomarkers and surrogate endpoints: preferred definitions and conceptual framework, *Clinical Pharmacology and Therapeutics*, 2001, **69**(3): 89–95.

[2] H Ledford, Drug markers questioned, *Nature*, 2008, **452**: 510–511.

[3] For examples of biomarker uses in these disciplines, see J Angerer, U Ewers and M Wilhelm, Human biomonitoring: state of the art, *International Journal of Hygiene and Environmental Health*, 2007, **210**: 201–228; P Duramad, IB Tager and NT Holland, Cytokines and other immunological biomarkers in children's environmental health studies, *Toxicology Letters*, 2007, **172**(1–2): 48–59.

What Should Biomarkers Predict?

A commonly discussed scenario in personalized medicine is the use of biomarkers to screen healthy individuals at regular intervals for early signs of disease onset.[4] It is important to note that such screening is not intended to test for every possible disorder in every individual. The number of false positives (and misses) will inevitably increase with the number of tests performed. Population-wide scans for large numbers of diseases would lead to large numbers of false positives with associated financial and personal costs.[5] Moreover, screening is only of use if early intervention has been shown to be beneficial, either in clinical terms (improved health outcomes), or at a personal level (e.g. preparing and planning for worsening health).

We will assume that regular screening will be used only to test for the onset of those disorders to which a person is at particularly high risk, and for which beneficial interventions exist. Disease risks could be calculated for each individual, given the individual's genetic and environmental exposures (discussed in the next two chapters). After considering the benefits of early detection, custom sets of biomarkers would then be used to regularly monitor the individual for signs of onset of those disorders to which the individual is considered particularly susceptible, and where early detection is deemed beneficial.

The Multiple Roles of Biomarkers in Personalized Medicine

Biomarkers differ in terms of their purpose, the technologies they employ, and the quantities they measure. To illustrate the multiple roles of biomarkers in personalized medicine, consider the following scenarios.

(i) A healthy individual is screened at regular intervals for early signs of a disease to which (s)he is deemed particularly susceptible. Such early detection screens must be carried out at repeated intervals on healthy individuals, so they must be non (or minimally) invasive, low-cost, and pose a negligible risk to the patient.

(ii) A screening test indicates early signs of a disorder. The patient will now undergo multiple diagnostic tests to verify and pinpoint the findings. Diagnostic biomarkers are used to assess disease severity, sub-type, and progression stage. Because they are used only as needed, and depending on the severity of the disorder, diagnostic tests can cost more, or be more invasive than screening tests. For example, more precise stratification

---

[4] See for example AD Weston and L Hood, Systems biology, proteomics, and the future of health care: toward predictive, preventative, and personalized medicine, *Journal of Proteome Research*, 2004, **3**(2): 179–196.

[5] See for instance E Marshall, A Bruising battle over lung scans, *Science*, 2008, **320**: 600–603.

of a disorder may be arrived at through invasive procedures that sample the affected tissues directly.

(iii) Following treatment, prognostic screening may be used to identify patients at risk of recurrence. For example the commercially available multiple-gene-expression tests Oncotype[6] and MammaPrint[7] estimate the probability of recurring breast cancer after initial treatment (e.g. surgery), and are commonly used to decide who should receive adjuvant therapy to avoid relapse.

(iv) For patients undergoing treatment, response biomarkers identify patient sub-types who will benefit from a particular treatment. For example, as noted in Chapter 1, a mutation in the *HER2* gene identifies breast cancer patients who will respond favorably to treatment with Herceptin. Response biomarkers are often developed by drug companies to identify patients who will benefit most from their drugs.

(v) Finally, kinetic biomarkers allow dosage selection, indicate toxicity, etc. for patients receiving a particular drug (or class of drugs). For example, the Roche Amplichip[8] tests for mutations in the Cytochrome P450 system to determine the rate at which an individual will metabolize a drug. It can be used to determine an appropriate dosage for anti-depressants and anti-psychotics, allowing patients to avoid lengthy trial and error approaches.

Imaging Anatomical and Physiological Biomarkers

Organ anatomy is a cumulative record of the interactions of genes with the environment. Anatomical imaging can reveal not only deformities (e.g. blocked arteries, neural degradation, or enlarged prostate), but also indicate the type (and therefore cause) and progression state (stage) of the disorder.

Physiological biomarkers such as heart rate, blood pressure, the smell of breath, the color of urine, the texture of the tongue, the knee reflex, and physical fitness tests have long been used for diagnosis. Imaging technologies extend such direct and indirect correlates of organ physiology by providing a broader range of measurable characteristics, greater detail and better discrimination capabilities.

In vivo imaging techniques have the advantage that there is little risk of pain, infection or other complications that can accompany surgical procedures. Moreover, the data is collected

---

[6] http://www.genomichealth.com/OncotypeDX/

[7] http://usa.agendia.com/en/mammaprint.html

[8] http://www.amplichip.us/

in the normal anatomical and functional context of cells and organs, and so is more representative of actual cellular state. Finally, because of their relatively low impact on patient well-being, some imaging and physiological measurements can be performed repeatedly to provide trend data (e.g. treatment/disease progression).

The most prominent current imaging technologies[9] are: ultrasound (US), X-ray computed tomography (CT), Magnetic Resonance Imaging (MRI), Positron Emission Tomography (PET), and various forms of visible-light optical imaging (OI). In essence, all imaging technologies can be divided into three categories:

- **transmission imaging** measuring the amount of incident energy (e.g. X-rays or visible light) that passes through the sample;
- **reflection imaging** measuring the proportion of reflected waves (e.g. echoes measured in ultrasound);
- **emission imaging**, where various agents are used to cause certain molecules or structures in the sample to emit energy (e.g. gamma rays, or fluorescence).

CT is distinguished from normal X-ray imaging in that it uses multiple cross-sectional images, and computational algorithms to deliver a three-dimensional anatomical image. Because of the need for multiple cross-sections, CT imaging typically involves about 100-fold more exposure to (harmful) ionizing radiation than a standard X-ray. This has tended to limit the use of CT imaging to cases where the benefit clearly outweighs the cumulative adverse effects (e.g. in cancer therapy, where CT is often used in combination with PET).

CT images can reveal not only anatomical features (e.g. tumors), but also correlates of dysregulated molecular pathways. For example, an analysis of CT images of liver tumors[10] found that image features such as texture heterogeneity and visible internal arteries can be correlated with specific gene expression patterns. Indeed, the authors were able to predict a 12-fold increased risk of microscopic venous invasion (indicating poor prognosis) using just two image features. The use of higher resolution imaging technologies could allow this approach to be generalized to other cancers.

---

[9] Reviewed in CH Contag, *In vivo* pathology: seeing with molecular specifcity and cellular resolution in the living body, *Annual Review of Pathology: Mechanisms of Disease*, 2007, **2**: 277–305; AR Kherlopian, A review of imaging techniques for systems biology, *BMC Systems Biology*, 2008, **2**: 74.

[10] E Segal *et al.*, Decoding global gene expression programs in liver cancer by noninvasive imaging, *Nature Biotechnology*, 2007, **25**(6): 675–680.

In **MRI**, high magnetic fields are used to align the subject's atomic nuclei to an applied magnetic field. A radio frequency pulse is used to briefly tilt the magnetized nuclei out of alignment with the magnetic field. Following the radio-pulse, the nuclei return to the magnetically aligned state on timescales that indicate the molecular environment surrounding the nuclei (i.e. the sample composition). The nuclei of the common forms of hydrogen, nitrogen, oxygen, sodium, phosphorous and potassium can all be used in this manner, while the nuclei of ^{13}C and ^{43}Ca isotopes can be used for dynamic tracing of tissue uptake.[11]

Contrast agents (usually containing metal ions) can improve MRI image quality by changing the number of fluctuating magnetic fields near nuclei. Anatomically localized (usually injected) contrast agents are often used to highlight features such as blood vessels, tumors, or inflammation. Unfortunately, most current MRI contrast agents have toxic effects. Moreover, in spite of its efficacy in numerous medical applications, the initial and operational costs of MRI have limited its uptake in some areas.[12]

Two recent extensions of MRI are taking it beyond anatomical imaging. Diffusion MRI visualizes flow, for example in blood vessels (MRI angiography) and neural axons. Dark–light shading is used to indicate the speed of movement, thus highlighting clinically important features such as occlusions.[13]

A magnetized atomic nucleus will absorb a radio wave at slightly different frequencies depending on the chemical structure of the molecule it is part of. This property is used in MRI spectroscopy to measure the presence of various molecules within a given region of interest. Superimposing MRI spectroscopy data onto anatomical MRI images can provide two and three-dimensional visualizations of the distributions of various analytes within the body.[14]

PET technology uses radio-labeled pharmacological agents to indirectly visualize the activity of specific molecular processes within the body. PET images are frequently superimposed on

---

[11] See for example K Golman, R in 't Zandt and M Thaning, Real-time metabolic imaging, *Proceedings of the National Academy of Sciences of USA*, 2006, **103**: 11270–11275.

[12] See for example WP Bandettini and AE Arai, Advances in clinical applications of cardiovascular magnetic resonance imaging, *Heart*, 2008, **94**: 1485–1495; JJ Nikken and GP Krestin, MRI of the kidney — state of the art, *European Radiology*, 2007, **17**: 2780–2793.

[13] See for example RR Moustafa and J-C Baron, Clinical review: imaging in ischaemic stroke — implications for acute management, *Critical Care*, 2007, **11**: 227.

[14] See for example E Casciani and GF Gualdi, Prostate cancer: value of magnetic resonance spectroscopy 3D chemical shift imaging, *Abdominal Imaging*, 2006, **31**: 490–499.

concurrently obtained CT scans to pinpoint the location of observed biochemical processes (e.g. high metabolic activity observed in some tumors).

Although PET technology has primarily been used to visualize metabolic processes so far, in principle it can be used to trace the activity of any molecular pathway *in vivo* as long as one or more molecules specific to that pathway can be radio-labeled with a positron-emitting isotope and delivered to the right cells.

The radio isotopes used in PET typically have half-lives in the order of half an hour. As a result, PET can only be performed at locations equipped with multi-million dollar cyclotron and radiochemistry laboratories. High costs and radiation exposure have so far limited the use of PET to analysis of severe conditions. Nevertheless, PET has found enormous success in the diagnosis, staging, and response-monitoring of many cancers,[15] as well as certain cardiological, pulmonary and neurological conditions.[16]

Single photon emission computed tomography (SPECT) measures gamma ray emissions from radionuclides directly. Compared to PET, SPECT radio compounds have much longer half-lives and do not need to be manufactured on-site immediately before use. As a result, SPECT imaging can be performed at much lower cost than PET. While the current resolution of SPECT images is fairly low (in the order of one millimeter), SPECT–CT combination imaging is highly effective in a number of specialist settings, such as the detection of sentinel lymph nodes in breast cancer.[17] Gating the acquisition of SPECT cardiac images by electrocardiogram features allows imaging of myocardial perfusion at specific points during the cardiac cycle.[18]

CT, MRI and PET imaging technologies will continue to be used widely in diagnosis, staging and treatment of severe disorders, as they are today. Indeed, their use is likely to become more widespread with reductions in toxicity and improvements in spatial resolution and the repertoire of available molecular targets. However, because of toxicity, cost, and the need for specialist operators and medical professionals, these technologies may not be suitable means of pre-symptomatic screening. The greatest impact of these technologies on

---

[15] N Sharma, D Neumann and R Macklis, The impact of functional imaging on radiation medicine, *Radiation Oncology*, 2008, **3**: 25.

[16] B Bybel *et al.*, PET and PET/CT imaging: what clinicians need to know, *Cleveland Clinic Journal of Medicine*, 2006, **73**(12): 1075–1087.

[17] IMC van der Ploeg *et al.*, The Hybrid SPECT/CT as an additional lymphatic mapping tool in patients with breast cancer, *World Journal of Surgery*, 2008, **32**(9): 1930–1934.

[18] G Germano and D Berman, *Clinical Gated Cardiac SPECT*, 2nd edition, Blackwell Futura, 2006.

personalized medicine (as opposed to medicine in general) may be pre-clinical: in research that characterizes disease processes, and in drug development.

In a similar vein, transmission, reflection and emission imaging using visible light (optical imaging) has shown great promise in animal studies,[9] but toxicity issues with the contrast agents may limit its near-term use to pre-clinical studies.

Ultrasound is essentially an evolution of the sonar technology used to detect submarines during the Second World War.[19] Reflections of waves in the frequency range 1–15 MHz are used to image a variety of lung[20] and heart[21] conditions, as well as fetal positioning, anatomy, heartbeat, fluid levels and placenta location during pregnancy. Recently, the use of microbubbles as contrast agents has greatly improved the resolution of ultrasound imaging. Moreover, microbubbles can be burst open in target tissues with focused ultrasound waves, thus providing an opportunity to deliver drugs, genes, or imaging probes locally.[22]

Ultrasound technology is safe and cheap. It offers millimeter-resolution images of structures up to a few centimeters inside the body. With recent improvements in image quality, ultrasound images have become easier to interpret, thus avoiding the need for specialist operators. Furthermore, the simplicity and low power requirements of ultrasound make it ideal as a bedside diagnostic tool. Indeed, there is already a "pocket" ultrasound scanner available commercially.[23]

Blood-Related Biomarkers

As we noted in Chapter 2, the cellular and liquid parts of blood offer a unique window onto the biochemical status of the entire body. Moreover, blood is easy to draw and is in plentiful supply, making it a convenient source of biomarkers.

---

[19] For a review of the theory, history and safety of ultrasound, see WD O'Brien, Ultrasound — biophysics mechanisms, *Progress in Biophysics and Molecular Biology*, 2007, 93(1–3): 212–255.

[20] B Bouhemad *et al.*, Clinical review: bedside lung ultrasound in critical care practice, *Critical Care*, 2007, 11: 205.

[21] Q Ciampi and Bruno Villari, Role of echocardiography in diagnosis and risk stratification in heart failure with left ventricular systolic dysfunction, *Cardiovascular Ultrasound*, 2007, 5: 34; M Galderisi, F Cattaneo and S Mondillo, Doppler echocardiography and myocardial dyssynchrony: a practical update of old and new ultrasound technologies, *Cardiovascular Ultrasound*, 2007, 5: 28.

[22] GA Husseini and WG Pitt, The use of ultrasound and micelles in cancer treatment, *Journal of Nanoscience and Nanotechnology*, 2008, 8(5): 2205–2215; K Ferrara, R Pollard and M Borden, Ultrasound microbubble contrast agents: fundamentals and application to gene and drug delivery, *Annual Review of Biomedical Engineering*, 2007, 9: 415–447.

[23] The Acuson P10 Pocket ultrasound system from Siemens: http://www.medical.siemens.com/

Immune cells. There are typically about 4,500 to 10,500 white blood cells per micro liter (1 mm^3) of blood. These circulating immune cells act as sentinels against infections and other adverse environmental effects for the entire body. They are also easy to collect and analyze at the DNA, RNA and protein levels. As such, they offer an excellent means of interrogating the state of the entire immune system, and detecting disease processes early on.[24] One complication is that white blood cells comprise at least half a dozen different cell types (neutrophils, eosinophils, basophils, B and T cells, monocytes, and macrophages). There are also considerable differences in the relative abundances of each cell type among healthy individuals.

Since different immune cells express different gene sets and perform very different functions, accurate characterization of biomarkers in circulating immune cells may require approaches that isolate and characterize particular cell types from whole blood. However, the very fact that there is considerable inter-individual variation in white blood cell counts can be used to identify cell-type specific genes in whole-blood transcriptome data.

The top panel in figure 7.1 shows the distribution of TH1 cells as a fraction of the total number of circulating CD4+ T cells measured in 16 healthy adults.[25] Note the ~10-fold variability among individuals.

The lower panel in figure 7.1 shows an example of a gene (*versican*) whose expression correlates with the concentration of a particular cell type (neutrophils) in individuals *and* is indicative of a disorder.[26] In this case, the disease is a rat model of Rheumatoid Arthritis.

The expression level of the gene identified (*versican*) is highly correlated with blood neutrophil concentration across a large number of rats, suggesting that *versican* expression in these blood samples is neutrophil-specific. This makes *versican* a less noisy biomarker than genes which are simultaneously expressed — and potentially differentially regulated — in multiple cell types. As may be expected of an autoimmune disease marker, Versican protein is associated with the regulation of inflammation.[27] Cell-type specific biomarkers for many other diseases are likely to be identified in similar ways in the near future.

---

[24] Proposed in PO Brown and L Hartwell, Genomics and human disease: variations on variation, *Nature Genetics*, 1998, **18**(2): 91–93.

[25] Figure from P Duramad *et al.*, Flow cytometric detection of intracellular Th1/Th2 cytokines using whole blood: validation of immunologic biomarker for use in epidemiologic studies, *Cancer Epidemiology, Biomarkers and Prevention*, 2004, **13**(9). Copyright 2004 by American Association for Cancer Research. Reproduced with permission.

[26] J Shou *et al.*, Identification of blood biomarkers of rheumatoid arthritis by transcript profiling of peripheral blood mononuclear cells from the rat collagen-induced arthritis model, *Arthritis Research and Therapy*, 2006, **8**: R28.

[27] TN Wight, Versican: a versatile extracellular matrix proteoglycan in cell biology, *Current Opinion in Cell Biology*, 2002, **14**(5): 617–623.

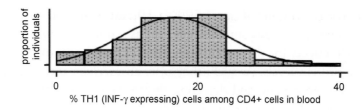

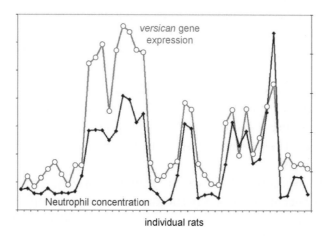

Figure 7.1: Inter-individual variations in immune cell composition can reveal biomarker genes.

Circulating Tumor Cells (CTCs). Tumors tend to shed cells into the blood stream. Many of these cells undergo apoptosis,[28] but some are thought to survive and contribute to metastatic cancer. The concentration of CTCs in blood was thought to be as low as five to ten cells per 7.5 ml of blood,[29] but recent measurements using a highly sensitive microfluidic platform suggest that estimate is about two orders of magnitude too low.[30] Indeed, for some cancers there may be well over 1,000 CTCs in 1 ml of whole blood.

Although current tests only detect CTCs in metastatic cancers, these improved measurements raise the promise of pre-metastasis CTC detection in the near future. Moreover, analysis of

---

[28] G Méhes *et al.*, Circulating breast cancer cells are frequently apoptotic, *American journal of Pathology*, 2001, **159**: 17–20.

[29] CL Sawyers, The cancer biomarker problem, *Nature*, 2008, **452**(7187): 548–552.

[30] S Nagrath *et al.*, Isolation of rare circulating tumour cells in cancer patients by microchip technology, *Nature*, 2007, **450**(7173): 1235–1239.

DNA, mRNA and protein content of captured CTCs could be used to establish the organ of origin, and the stage of development of the cancer.[31]

In a recent study involving 27 patients with metastatic non-small cell lung cancer,[32] microfluidic capture of CTCs found a median of 74 cells per milliliter of blood. CTC DNA was isolated from 10 ml of blood, and genotyped to identify patients who acquire mutations that confer resistance to some treatments.

Mutations acquired during or after treatment appear to occur in a variety of cancers. Since microfluidic CTC genotyping requires only blood samples, it can be repeated at frequent intervals to identify such cases.

In the above, study, only the capture of CTCs was performed using microfluidics. However, microfluidic devices that receive whole-blood samples, separate the cells from plasma and perform allele-specific on-chip PCR have already been demonstrated.[33] Thus it seems likely that within the next five years, we will see inexpensive, easy-to-use "lab-on-chip" commercial devices for blood-cell genotyping.

A reduction in the number of plasma CTCs is associated with a positive response to therapy, while an increase in the number of CTCs indicates with tumor progression.[34] Thus, accurate counting of CTCs may in itself be a useful means of monitoring responses to cancer treatment. A high-speed, low-cost microfluidic platform for on-chip counting of CTCs in milliliter quantities of whole blood was recently demonstrated to recover nearly all CTCs with high purity down to 10 CTCs per milliliter.[35] Thus low-cost, routine CTC counting may be available clinically in the near future.

Plasma proteins. The plasma proteome is unique in that it does not represent a particular cellular proteome, but instead reflects the collective proteome of all cells in the body. Apart from constitutive proteins, at least three mechanisms shed cellular proteins into the blood-stream. First, signaling molecules such as hormones and cytokines are routinely secreted into

---

[31] JW Uhr, One-stop shop, *Nature*, 2007, **450**(7173): 1168–1169.

[32] S Maheswaran *et al.*, Detection of mutations in EGFR in circulating lung-cancer cells, *New England Journal of Medicine*, 2008, **359**(4): 366–377.

[33] CJ Easley *et al.*, A fully integrated microfluidic genetic analysis system with sample-in–answer-out capability, *Proceedings of the National Academy of Science of USA*, 2006, **103**(51): 19272–19277.

[34] M Cristofanilli *et al.*, Circulating tumor cells, disease progression, and survival in metastatic breast cancer, *New England Journal of Medicine*, 2004, **351**: 781–791.

[35] AA Adams *et al.*, Highly efficient circulating tumor cell isolation from whole blood and label-free enumeration using polymer-based microfluidics with an integrated conductivity sensor, *Journal of the American Chemical Society*, 2008, **130**(27): 8633–8641.

the plasma. Second, cell surface proteins may be released into the plasma, particularly after undergoing cleavage. Third, cell damage or death will release some of the cellular content into the plasma. Thus, blood plasma proteins offer a unique means of probing the state of the entire body. On the other hand, the complexity of the plasma proteome makes its analysis particularly challenging.

To date, about 3,800 proteins have been identified in human plasma.[36] Although each of these proteins is thought to be present in dozens of modified states, plasma protein modification states and isoforms have so far not been mapped.

As summarized in figure 7.2,[37] constitutive or classical plasma proteins are several orders of magnitude more abundant than signaling molecules and proteins that "leak" into the blood stream from various tissues. The last two groups include the most likely candidates for disease biomarkers, as indicated by the distribution of lightly shaded disks in the plot below. The challenge of identifying biomarker proteins among the constitutive proteins, which may be 10–12 orders of magnitude more abundant, is truly like looking for a needle in a haystack.

About 3000 plasma proteins were identified by the Human Plasma Proteome Project (HPPP) in 2005.[38] In figure 7.2, note how the HPPP findings are dominated by constitutive proteins. This is not surprising, given that the 22 most abundant classical plasma proteins constitute more than 99% of the total mass.[39] To overcome this challenge, today most analyses of plasma proteome start by separating and removing high-abundance, constitutive plasma proteins.[39]

Research in plasma proteomics is still in its infancy, and mass-spectrometry (MS) technology is evolving rapidly. Thus, the current catalogue of plasma proteins should be seen as more illustrative than definitive. For example, the HIP^2 database,[40] which brings together

---

[36] The Plasma Proteome Database (http://www.plasmaproteomedatabase.org/) listed 3,778 proteins in Dec 2008; see also B Muthusamy *et al.*, Plasma Proteome Database as a resource for proteomics research, *Proteomics*, 2005, **13**: 3531–3536.

[37] This figure is reproduced by kind permission of Dr. Ralph Schiess, ETH Zurich. Published as part of RF Service, Will biomarkers take off at last? *Science*, 2008, **321**: 1760; see also NL Anderson and NG Anderson, The human plasma proteome: history, character, and diagnostic prospects, *Molecular and Cellular Proteomics*, 2002, **1**: 845–867.

[38] HPPP website is http://www.hupo.org/research/hppp/; the list of the proteins identified by HPPP in 2005 is available at http://www.peptideatlas.org/hupo/hppp/. Only 889 of the 3,020 HPPP identifications have >95% confidence. See DJ States *et al.*, Challenges in deriving high-confidence protein identifications from data gathered by a HUPO plasma proteome collaborative study, *Nature Biotechnology*, 2006, **24**(3): 333–338. A list of this high-confidence set is given at the preceding URL.

[39] W-J Qian *et al.*, Enhanced detection of low abundance human plasma proteins using a tandem IgY12-SuperMix immunoaffinity separation strategy, *Molecular and Cellular Proteomics*, 2008, **7**: 1963–1973.

[40] http://discern.uits.iu.edu:8340/HIP2/

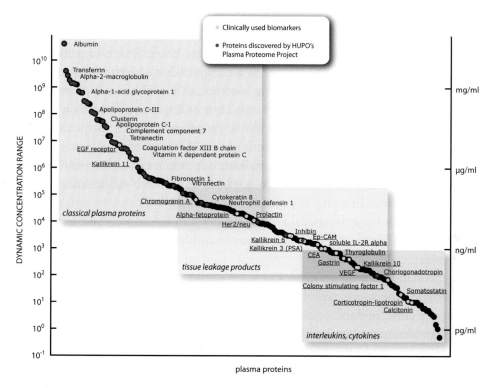

Figure 7.2: Concentration ranges of the plasma proteome and clinical blood biomarkers.

protein identifications from four different sources and eight different MS platforms, currently lists 12,787 candidate plasma proteins. But less than one percent of the proteins listed in HIP[2] were identified by all four groups contributing data.[41]

Between 86% and 94% of human genes are alternatively spliced into two or more protein isoforms.[42] Many current methods measure only total protein abundance. However, since different isoforms frequently perform different cellular functions, clinical protein biomarkers will likely need to be isoform-specific. Moreover, protein activity levels commonly depend on chemical post-translational modifications such as phosphorylation. Thus, true measures of the activity of a particular protein isoform will also need to be modification-specific.

Other challenges in plasma proteomics include large sample volumes, complex sample preparation, and the high cost of mass spectrometry. However, it is important to make a

---

[41] S Saha *et al.*, HIP2: an online database of human plasma proteins from healthy individuals, *BMC Medical Genomics*, 2008, **1**: 12.

[42] ET Wang *et al.*, Alternative isoform regulation in human tissue transcriptomes, *Nature*, 2008, **456**: 470–476.

distinction between the discovery of disease-associated biomarkers in plasma, and clinical use of validated biomarkers for diagnosis. The former can be slow, costly and complex as long as it successfully delivers medically useful biomarkers. The latter, i.e. testing for a set of well-characterized biomarkers in a blood sample, is a considerably easier task.

Tests for specific plasma biomarkers can use much simpler sample preparation, smaller sample quantities, and more cost-effective assays. For example, a microfluidic antibody array that automatically isolates the plasma from whole blood, and then reports antibody binding levels as a fluorescent signal was recently demonstrated to perform an entire blood test for little more than the cost of the antibodies, in about ten minutes.[43] Moreover, the device was able to identify low-abundance cytokines (e.g. IL-6) as well as a tissue leakage protein: the (partial) cancer biomarker Prostate Specific Antigen (PSA).[44]

Increasing evidence suggests that there is a humoral immune response to most tumors.[45] Thus, fast, low-cost microfluidic early detection of antibodies to multiple cancer types may soon be possible.

Signaling proteins secreted into plasma by the immune system are considerably rarer than PSA. For example, there are typically fewer than a hundred thousand molecules of the cytokine IL-6 per cubic millimeter of blood.[46] The ability of the above microfluidic device to detect such a low abundance molecule suggests that it may soon be possible to infer the state of the immune system from the plasma cytokine profile.

The above device uses fluorescence microscopy for signal detection. An integrated, low-cost, microelectronic/microfluidic systems capable of quantitative electronic measurements of antibody binding in microliter quantities of blood was recently demonstrated.[47] Thus, future plasma proteomic measurements are likely to be portable, low-cost, and easy to use.

---

[43] R Fan *et al.*, Integrated barcode chips for rapid, multiplexed analysis of proteins in microliter quantities of blood, *Nature Biotechnology*, 2008, **26**(12): 1373–1378.

[44] PSA has a molecular weight of 33KDa, i.e. 33Kg of PSA contains ~6×10^{23} (Avogadro's number) PSA molecules. PSA is present in plasma at concentrations of up to 10 ng/ml (older individuals tend to express more PSA). This corresponds to up to ~200 million molecules of PSA in one 1 mm^3 (roughly one drop) of blood.

[45] J Qiu, Occurrence of autoantibodies to Annexin I, 14-3-3 theta and LAMR1 in prediagnostic lung cancer sera, *Journal of Clinical Oncology*, 2008, **26**(31): 5060–5066. For a review of the topic, see KS Anderson and J LaBaer, The sentinel within: exploiting the immune system for cancer biomarkers, *Journal of Proteome Research*, 2005, **4**(4): 1123–1133.

[46] IL-6 has a molecular weight of ~24KDa and has a plasma concentration of about 1–10 pg/ml (from Anderson and Anderson, 2002, Ref. 37).

[47] H Lee *et al.*, Chip–NMR biosensor for detection and molecular analysis of cells, *Nature Medicine*, 2008, **14**(8): 869–874.

The fact that microfluidic plasma protein analysis requires only a single drop of blood allows frequent and routine sample collection. This has two desirable effects. Firstly, patient health status can be determined from trends over time, rather than single assays (which suffer from intra- and inter-patient variabilities). Secondly, because the assays are molecularly detailed and quantitative, retrospective data mining of patient records can identify patterns not apparent at the outset. This leads to a virtuous cycle: better monitoring technology leads to better understanding of the causes of disorders, which leads to the identification of better biomarkers and treatments, and so on.

Saliva-Based Biomarkers

Whole saliva contains about 1 mg/ml of total protein.[48] Three different oral glands contribute saliva,[49] and secretions from each gland have a distinctly different composition.[50] In addition to these secretions, there appears to be considerable leakage from the blood and interstitial fluid into saliva. Indeed, whole saliva includes exfoliated cells and cellular fragments, as well as "leakage" from the oral bacteriome and nasal and bronchial processes.

Whole saliva genomic DNA is routinely used for genetic testing[51] since a simple mouthwash typically yields about 10–100 micrograms of human genomic DNA.[52] Moreover, the cell-free portion of saliva has been found to contain fragments of more than 850 different RNA transcripts, including many mRNAs.[53]

Many saliva proteins are also found in plasma, and a number of the salivary proteins map to pathways involved in neurodegenerative conditions (e.g. Alzheimer's, Huntington's, and Parkinson's), cancers (breast, colorectal, and pancreatic), and diabetes.[50] Although members of just seven protein families make up more than 90% of the protein content of whole

---

[48] S Hu *et al.*, Salivary proteomics for oral cancer biomarker discovery, *Clinical Cancer Research*, 2008, **14**(19): 6246–6252.

[49] DT Wong, Salivary diagnostics, *American Scientist*, 2008, **96**(1): 37–43.

[50] P Denny *et al.*, The proteomes of human parotid and submandibular/sublingual gland salivas collected as the ductal secretions, *Journal of Proteome Research*, 2008, 7(5): 1994–2006.

[51] The stability of DNA also makes sample preparation and storage straightforward; see for example EA Ehli *et al.*, Using a commercially available DNA extraction kit to obtain high quality human genomic DNA suitable for PCR and genotyping from 11-year-old saliva saturated cotton spit wads, *BMC Research Notes*, 2008, 1: 133.

[52] A Lum and L le Marchand, A simple mouthwash method for obtaining genomic DNA in molecular epidemiological studies, *Cancer Epidemiology, Biomarkers and Prevention*, 1998, 7: 719–724.

[53] Z Hu *et al.*, Exon-level expression profiling: a comprehensive transcriptome analysis of oral fluids, *Clinical Chemistry*, 2008, **54**(5): 824–832.

saliva,[54] more than 1,400 other proteins have been identified in saliva (756 with high confidence).[55]

These observations have raised hopes that saliva biomarkers may provide a convenient, non-invasive means of detecting a wide range of disorders. Indeed, a commercial saliva test for HIV 1 and 2 antibodies with >99% sensitivity and >99.5% specificity has been available in the US for a number of years.[56]

In a preliminary study, oral cancer could be detected using a five-protein signature in saliva.[48] Remarkably, a recent small-scale study of saliva proteins in women with breast cancer found 49 proteins that could collectively distinguish between healthy women, women with benign tumors, and women with carcinomas.[57]

Although the concentration of solutes in saliva is typically much lower than blood, integrated microfluidic systems that use microliter to milliliter quantities of saliva and can deliver rapid, in-field, low-cost and non-invasive diagnostics have been demonstrated for both protein detection[58] and monitoring of small molecules (e.g. drugs).[59] These systems also include on-chip filters for sample enrichment, thus performing an important sample preparation step automatically.

For protein and RNA assays, collection and preparation of saliva must be carried out carefully to ensure reproducibility. However, the protocols and devices used to collect the various types of saliva are straightforward, cheap, and without adverse effects.[49] Moreover, multi-milliliter quantities of saliva can be collected in a single sitting.

Biomarkers in Urine and Feces

Because of their natural and abundant availability, urine and feces are attractive media for health monitoring, screening, and diagnostic assays.

---

[54] I Messana *et al.*, Facts and artifacts in proteomics of body fluids. What proteomics of saliva is telling us? [*sic*] *Journal of Separation Science*, 2008, **31**(11): 1948–1963.

[55] W Yan *et al.*, Systematic comparison of the human saliva and plasma proteomes, *Proteomics — Clinical Applications*, 2009, **3**(1): 116–134.

[56] KP Delaney *et al.*, Performance of an oral fluid rapid HIV-1/2 test: experience from four CDC studies, *Aids*, 2006, **20**(12): 1655–1660.

[57] CF Streckfus *et al.*, Breast cancer related proteins are present in saliva and are modulated secondary to ductal carcinoma *in situ* of the breast, *Cancer Investigation*, 2008, **26**(2): 159–167.

[58] AE Herr *et al.*, Microfluidic immunoassays as rapid saliva-based clinical diagnostics, *Proceedings of the National Academy of Sciences of USA*, 2007, **104**(13): 5268–5273.

[59] E Fu *et al.*, SPR imaging-based salivary diagnostics system for the detection of small molecule analytes, *Annals of the New York Academy of Sciences*, 2007, **1098**: 335–344.

Healthy adults excrete in the region of 1.5 liters of urine every 24 hours. Although ultra-filtration in the kidneys retains the vast majority of large molecules, significant amounts of DNA, RNA, and proteins (or protein fragments) from the circulatory system can be detected in urine.

A healthy person's urine typically only contains relatively small amounts of DNA, RNA and proteins (e.g. less than 100 mg/l of proteins), but many disorders lead to elevated levels of specific molecular species. For example, acute kidney injury has been shown to be detectable by the presence of the transcription factor ATF3 in urine,[60] and prostate cancer can be detected from the urine levels of PCA3, a prostate-specific noncoding mRNA found in more than 95% of prostate cancer specimens.[61] Also, a variety of disorders can lead to excessively high total protein concentration in urine (proteinuria), which can be detected with a simple dipstick test.[62]

Several urine tests can identify bladder cancer. NMP22, a protein biomarker of bladder cancer found in urine, can be assayed with a simple commercially available point-of-care device (http://www.matritech.com/). A study of 103 patients found that while the NMP22 test alone is not sufficiently sensitive (<50%), 99% of malignancies are detected when NMP22 testing is combined with cytoscopy.[63] Another commercially available product (http://www.urovysion.com/), detects aneuploidy of chromosomes 3, 7 and 17, and loss of a locus on chromosome 9 using genomic DNA extracted from urine samples of bladder cancer candidates.

In addition to aneuploid DNA from bladder cancer, DNA from colorectal cancer cells has been shown to be present in urine. Remarkably, detection of a mutation-bearing DNA fragment associated with colorectal cancer has been shown to be more reliable in urine than in serum[64] (because urine has lower overall concentrations of confounding macromolecules than serum, and so is easier to process).

---

[60] H Zhou *et al.*, Urinary exosomal transcription factors, a new class of biomarkers for renal disease, *Kidney International*, 2008, **74**(5): 613–621.

[61] MPMQ van Gils *et al.*, The time-resolved fluorescence-based PCA3 test on urinary sediments after digital rectal examination; a Dutch multicenter validation of the diagnostic performance, Clinical Cancer Research, 2007, **13**: 939–943.

[62] JA Simerville *et al.*, Urinalysis: a comprehensive review, *American Family Physician*, 2005, **71**: 1153–1162.

[63] HB Grossman *et al.*, Surveillance for recurrent bladder cancer using a point-of-care proteomic assay, *Journal of the American Medical Association*, 2006, **295**: 299–305.

[64] Y-H Su *et al.*, Detection of mutated K-*ras* DNA in urine, plasma, and serum of patients with colorectal carcinoma or adenomatous polyps, *Annals of the New York Academy of Sciences*, 2008, **1137**: 197–206.

Excreted metabolites in urine are widely used to detect environmental toxins,[65] and also to detect metabolic disorders. But urine metabolite composition is also affected by complex disorders. For example, a comparison of about 700 urine metabolites in Parkinson's Disease (PD) sufferers and controls found that the levels of several metabolites were distinctly different in the two groups. Urea levels were generally lower in PD, while suberic acid levels were generally higher.[66] No single metabolite was an accurate biomarker for PD, but supervised classification of 700 metabolite readings clearly separated the PD and control groups, as shown in figure 7.3[67] (reproduced from Ref. 66).

The figure shows the metabolite readings of the two groups projected into the space of responses by the three best linear discriminants separating the PD and control groups. The discs represent individuals in the PD population. The squares represent the healthy controls. The axis closest to horizontal corresponds to the value of the first discriminant. The PD and

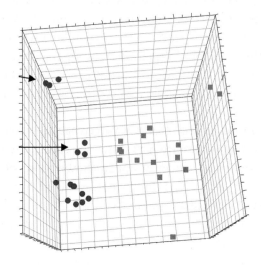

Figure 7.3:   Multiparameter biomarker-based discrimination of Parkinson's disease.

---

[65] See for example M-C Fortin, G Carrier and M Bouchard, Concentrations versus amounts of biomarkers in urine: a comparison of approaches to assess pyrethroid exposure, *Environmental Health*, 2008, 7: 55.

[66] AW Michell *et al.*, Metabolomic analysis of urine and serum in Parkinson's disease, *Metabolomics*, 2008, 4: 191–201.

[67] Figure reproduced with kind permission from Springer Science and Business Media: figure 2 of AW Michell *et al.*, Metabolomic analysis of urine and serum in Parkinson's disease, *Metabolomics*, 2008 4: 191–201.

control populations are clearly separate from each other along this axis, suggesting that a multi-metabolite signature can correctly diagnose PD patients.

The second and third discriminant axes reveal another feature of the urine metabolite profiles: the PD populations fall into three distinct groups (the main group at the bottom, plus two smaller groups indicated by arrows). Thus, using a multi-metabolite signature it may be possible to sub-type or stage PD patients based on urine samples alone.

The above findings are based on a small study (23 PD patients, 23 controls) and would need to be verified in much larger trials before clinical consideration. But in terms of our discussion here, these results provide a good example of how non-invasive assays and multi-marker signatures may be used for early disease detection, sub-typing and staging.

Urine composition appears to vary with genetic background. A comparison of 146 African-American post-menopausal women with 330 Caucasian-American counterparts identified significant differences in several key analytes, including creatinine, calcium, potassium, and magnesium.[68] There were also differences in total volume and pH. Thus, it may be beneficial, and possibly necessary, to tailor urine biomarker significance thresholds to specific genetic populations.

Due to variable degradation rates, urine tests typically have to be performed within one to two hours of sample collection. Kits that allow self-testing by patients can thus reduce costs and enable routine testing. For example, in Japan, the Tanita Corporation (http://www.tanita.com/en/) introduced a hand-held urine glucose monitor in 2008 (model UG-210, not available outside Japan). A very low-cost paper-based microfluidic device that can measure glucose and total protein levels in urine was recently demonstrated.[69] The device requires no special equipment or handling and may therefore prove ideal in remote and under-developed regions.

We noted in Chapter 4 that the human gastro-intestinal (GI) microbiome comprises hundreds of distinct bacterial populations, and that the composition of this population varies with race, gender, diet and other factors. A recent analysis of the microbiome of seven (Chinese) individuals and their urine metabolic profiles suggests that specific commensal microbial groups may play distinct metabolic roles.[70] Thus, drug uptake efficiency and

---

[68] EN Taylor and GC Curhan, Differences in 24-hour urine composition between black and white women, *Journal of the American Society of Nephrology*, 2007, **18**: 654–659.

[69] AW Martinez *et al.*, Simple telemedicine for developing regions: camera phones and paper-based microfluidic devices for real-time, off-site diagnosis, *Analytical Chemistry*, 2008, **80**(10): 3699–3707.

[70] M Li *et al.*, Symbiotic gut microbes modulate human metabolic phenotypes, *Proceedings of the National Academy of USA*, 2008, **105**(6): 2117–2122.

onset/progression of many metabolism-related disorders may depend significantly on the state of an individual's GI microbiome.

Tests detecting blood in the feces are already commonly used for a variety of indications. Many of these tests can be performed at home by the patient using commercially available, low-cost and disposable kits.[71] Recently developed stool DNA tests for early diagnosis of specific disorders such as colorectal cancer show great promise.[72] As the specific roles of commensal bacterial groups become clear, fecal bacterial DNA tests can be used to detect metabolic susceptibilities and dysregulated enteric processes.

Stool DNA and proteomic tests can also be used to sub-type GI infections. For example, proteomic analysis of stool samples from 32 patients with cholera identified 909 Vibrio cholerae proteins, of which 25 were related to pathogenesis and 15 were immunogenic.[73] In another recent study, sequencing of viral DNA in the stool of 12 children with diarrhea allowed identification of the viruses responsible.[74] Notably, five of the children (~40%) were infected with more than one virus, and a number of novel viruses were detected. The identification of feces-borne pathogen-specific biomarkers allows the development of highly specific, rapid and low-cost tests.[75]

Beyond viruses and bacteria, a recent animal study suggests that infection by prion proteins can also be detected in feces long before the animals become symptomatic.[76] Stool analysis will likely reveal many more early diagnostic biomarkers in the coming years.

Skin, Breath, Sweat and Tears

Exhaled breath contains a variety variety of organic compounds, and micron-sized water droplets carrying tissue leakage proteins such as cytokines.[77] While the protein content of

---

[71] See for example http://www.helena.com/occultscreen.htm.

[72] DA Ahlquist *et al.*, Stool DNA and occult blood testing for screen detection of colorectal neoplasia, *Annals of Internal Medicine*, 2008, **149**: 441–450.

[73] RC LaRocque *et al.*, Proteomic analysis of vibrio cholerae in human stool, *Infection and Immunity*, 2008, 6(9): 4145–4151.

[74] SR Finkbeiner *et al.*, Metagenomic analysis of human diarrhea: viral detection and discovery, *PLoS Pathogens*, 2008, **4**(2): e1000011.

[75] See for example W Yamazaki *et al.*, Sensitive and rapid detection of cholera toxin-producing Vibrio cholerae using a loop-mediated isothermal amplification, *BMC Microbiology*, 2008, **8**: 94.

[76] JG Safar *et al.*, Transmission and detection of prions in feces, *Journal of Infectious Diseases*, 2008, **198**(1): 81–89.

[77] Reviewed in DH Conrad, J Goyette and PS Thomas, Proteomics as a method for early detection of cancer: a review of proteomics, exhaled breath condensate, and lung cancer screening, *Journal of General Internal Medicine*, 2007, **23**(Suppl 1): 78–84; J Hunt, Exhaled breath condensate — an overview, *Immunology and Allergy Clinics of North America*, 2007, **27**(4): 587–596.

breath remains to be characterized, picomolar concentrations of some 200 volatile organic compounds have been detected in human breath.

The abundance of the enzyme that normally metabolizes these compounds is up-regulated in lung cancer, such that it is possible to screen for lung cancer using a low-cost nine-compound breath test.[78] In this way, the adverse effects and cost of CT scans and biopsies can be avoided in all but those who test positive with a simple, low-cost and non-invasive breath screen.

Cystic Fibrosis (CF) arises from a mutation in the CF transmembrane conductance regulator protein, resulting in defective transport of chloride through cell membranes and defective mucociliary action. As a result, CF is often accompanied by chronic airway infection, inflammation and oxidative stress. Picogram per milliliter concentrations of interferon gamma (IFN-γ) and a number of other inflammatory biomarkers can be detected in the breath condensates of CF patients. In a sample of 48 CF children and 50 controls, a panel of three molecular biomarkers in breath condensates could identify CF patients with about 80% sensitivity and specificity.[79] Thus, it may soon be possible to screen for CF (and rate its severity) using a completely non-invasive breath test.

The human **tear** performs multiple maintenance functions and as a result contains many of the same molecules as serum, including a variety of proteins, lipids, and small molecules (e.g. glucose, lactate and urea).[80] Tear composition is also known to be affected by environmental factors such as smoking, and a number of disorders.[81] Consequently, tears offer a potentially non-invasive window onto aspects of health.

The challenges posed by tear biomarkers and the technologies developed to address them offer a good example of the shape of things to come in personalized health monitoring.

Tear composition depends on the stimulus (e.g. mechanical pressure, reflex to a flash of light, chemicals) and can vary dramatically, as illustrated in figure 7.4.[82] Shown are the tear glucose levels in a single individual measured hourly over a five-day period. The large peak around Day 2, 5–6 pm is due to measurements following swimming. In a similar vein, onion-stimulated tears have been found to have seven to eight-fold higher concentrations

---

[78] M Phillips *et al.*, Detection of lung cancer with volatile markers in the breath, *Chest*, 2003, **123**: 2115–2123.

[79] CMHHT Robroeks *et al.*, Biomarkers in exhaled breath condensate indicate presence and severity of cystic ?brosis in children, *Pediatric Allergy and Immunology*, 2008, **19**: 652–659.

[80] Reviewed in JM Tiffany, Tears in health and disease, *Eye*, 2003, **17**: 923–926.

[81] See for example GRC Baker *et al.*, Altered tear composition in smokers and patients with Graves Ophthalmopathy, *Archives of Ophthalmology*, 2006, **124**: 1451–1456.

[82] KM Daum and RM Hill, Human tear glucose, *Investigative Ophthalmology and Visual Science*, 1982, **22**: 509–514. Copyright 1982 by Investigative Ophthalmology and Visual Science. Reproduced with permission.

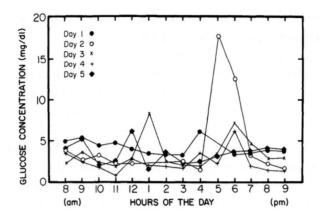

Figure 7.4:   Tear glucose concentration in a single individual measured over five days.

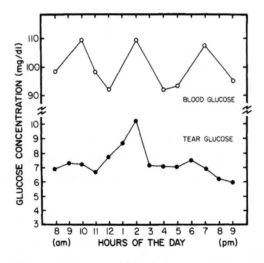

Figure 7.5:   Characteristic tear and blood glucose variations during the day.

of glucose.[83] Thus, environmental factors can have dramatic affects on tear analyte readings.

In addition to environmental stimuli, the diurnal cycle has a significant effect on tear composition, as illustrated in figure 7.5.[82] Shown are blood and tear glucose levels averaged over multiple individuals and days. Although the glucose diurnal cycles in blood and tear are

---

[83] CR Taormina *et al.*, Analysis of tear glucose concentration with electrospray ionization mass spectrometry, *Journal of the American Society for Mass Spectrometry*, 2007, **18**(2): 332–336.

clearly highly correlated, the upswing after lunch appears to be much more pronounced in tear than in blood. Thus, tear glucose levels measured at different times of day may need to be scaled differently to reflect blood glucose levels.

Therefore, it may be necessary to monitor tear biomarkers over multiple samples collected during the day. In turn, the need for repeated daily measurements raises a requirement for assays that are non-invasive, low-cost, rapid and simple to perform. A promising solution is offered by photonic crystals that can sense various analytes at physiological temperature, ionic strength and pH. The crystals are embedded in hydrogels which change volume as the analyte concentration varies. This changes the crystal lattice spacing, which changes the wavelength of diffracted light. The system has been shown to be able to detect tear glucose levels in the full physiological range within minutes,[84] leading to the futuristic proposal of contact lenses that change color depending on tear glucose levels.[85]

The **skin** provides an easy to monitor source of live cells in their natural context. In addition to dermatology-specific tests (e.g. for melanomas, skin allergies and infections), a number of non-dermatological disorders can be diagnosed with skin tests. The best established of these is probably the Mantoux tuberculin skin test, which measures the readiness of the adaptive immune system to respond to a tuberculosis infection. The Mantoux test has been used for over a century and continues to be one of the most effective and cost-effective means of tuberculosis testing.[86]

A colorimetric assay of skin cholesterol levels was demonstrated over 50 years ago.[87] Since then, studies have established that skin cholesterol levels correlate with rates of cardiovascular disease.[88] Simple, rapid and non-invasive colorimetric skin cholesterol tests have become available for both home and clinical use.[89]

---

[84] M Ben-Moshe, VL Alexeev and SA Asher, Fast responsive crystalline colloidal array photonic crystal glucose sensors, *Analytical Chemistry*, 2006, **78**: 5149–5157.

[85] VL Alexeev *et al.*, Photonic crystal glucose-sensing material for noninvasive monitoring of glucose in tear fluid, *Clinical Chemistry*, 2004, **50**(12): 2353–2360.

[86] A Nienhaus, A Schablon and R Diel, Interferon-Gamma release assay for the diagnosis of latent TB infection — analysis of discordant results, when compared to the tuberculin skin test, *PLoS ONE*, **3**(7): e2665; BA Winje *et al.*, School based screening for tuberculosis infection in Norway: comparison of positive tuberculin skin test with interferon-gamma release assay, *BMC Infectious Diseases*, 2008, **8**: 140.

[87] PR Moore and CA Baumann, Skin sterols. I. Colorimetric determination of cholesterol and other sterols in skin, *Journal of Biological Chemistry*, 1952, **195**(2): 615–621.

[88] DL Sprecher *et al.*, Skin tissue cholesterol (SkinTc) is related to angiographically-defined cardiovascular disease, *Atherosclerosis*, 2003, **171**: 255–258.

[89] See for example the PREVU Skin Sterol test kit http://www.premdinc.com/products_prevu.htm.

Because skin cells have many structures and processes in common with other body cells, many complex disorders have detectable side effects in the skin. For example, diabetes mellitus is often accompanied by a variety of skin conditions, some of which (e.g. skin tags and Candida infections) can be early (pre-symptomatic) indicators of diabetes.[90]

Genetic disorders affecting cell–cell contacts can lead to complex, multi-factor pathologies such as Charcot–Marie–Tooth disease, deafness, and arrhythmias. When caused by cell contact defects, these conditions are often accompanied by skin disorders that may act as early warning signals.[91]

Because it is on the surface of the body, skin tissue can be inspected using direct optical imaging. Hand held dermatoscopes[92] (essentially magnifiers with a light source and a liquid medium between the viewing lens and the skin) are widely used to enhance visual inspections of skin. Non-invasive near-infrared and visible-light imaging systems can visualize subcutaneous vasculature and blood oxygenation levels without using contrast agents or radiation.[93] Using contrast agents allows visualization of molecular biomarkers in the skin.[9] This may be useful for treatment monitoring since drug uptake and clearance rates in the skin can be very different from those in the blood.[94]

Human **sweat** is primarily water, but it also contains a diverse array of organic and inorganic molecules. As discussed in Chapter 6, cystic fibrosis is easily detected by measuring sweat chloride levels. The concentration of sweat solutes varies considerably from individual to individual, across body-parts,[95] and according to the cause of sweating (e.g. exercise versus heat). MRI spectroscopy analysis of heat-induced sweat in young, healthy men and women (30 of each) found no significant differences between men and women. However, while only lactate

[90] S van Hattem, AH Bootsma and HB Thio, Skin manifestations of diabetes, *Cleveland Clinic Journal of Medicine*, 2008, **5**(11): 772–787.

[91] JE Lai-Cheong, K Arita and JA McGrath, Genetic diseases of junctions, *Journal of Investigative Dermatology*, 2007, **127**: 2713–2725.

[92] See http://www.dermatoscopes.com.

[93] FP Wieringa *et al.*, Remote non-invasive stereoscopic imaging of blood vessels: first *in-vivo* results of a new multispectral contrast enhancement technology, *Annals of Biomedical Engineering*, 2006, **34**(12): 1870–1878; A Vogel *et al.*, Using noninvasive multispectral imaging to quantitatively assess tissue vasculature, *Journal of Biomedical Optics*, 2007, **12**(5): 051604.

[94] For example, for antihistamines used to treat urticaria. See I Jáuregui *et al.*, Antihistamines in the treatment of chronic urticaria, *Journal of Investigational Allergology and Clinical Immunology*, 2007, **17**(Suppl 2): 41–52.

[95] MJ Patterson, SDR Galloway and MA Nimmo, Variations in regional sweat composition in normal human males, *Experimental Physiology*, 2000, **85**(6): 869–875.

could be detected in the sweat collected from some individuals, the sweat of others included as many as 100 or more metabolites[96] (including electrolytes, amino acids and lipids).

Ten proteins have so far been characterized in sweat.[97] Most are associated with skin surface antimicrobial activity and response to inflammation. Serum albumin is also present in sweat. However, analyte concentrations in sweat do not always reflect the state of blood. For example, it has been known for several decades that sweat urea (but not uric acid) levels are typically 2–3 times higher than plasma concentrations, and that at least some of sweat urea comes from other sources.[98] Thus, sweat analysis may provide complementary information to blood biomarkers.

Microbial populations residing on the skin interact with sweat. Nearly 5000 different volatile compounds have been detected in sweat,[99] though it is not clear at present what proportion are intrinsic to sweat and what proportion are due to the processing of sweat by the skin microbiome. Remarkably, there appears to be little commonalty between individuals, suggesting that sweat odors are highly person-specific and may prove to be useful biomarkers.

Challenges in Biomarker Development

Biomarkers are usually selected on the basis of observed correlation between the marker and the biological process of interest. Correlation is a statistical measure and does not imply causality. Apart from the molecule(s) directly, mechanistically, and causally underlying a disorder, all other medical biomarkers are merely statistical correlates of the diseases of interest. Thus, biomarkers cannot be expected to be perfect predictors. On the other hand, evaluating the performance of biomarkers can be very expensive. Recent debate about the value of screening for lung cancer using CT scans[100] offers an illustrative example.

There are about 45 million smokers in the USA, all of whom can be considered at risk of lung cancer, and therefore candidates for screening. Each scan costs a few hundred US dollars,

---

[96] M Harker *et al.*, Study of metabolite composition of eccrine sweat from healthy male and female human subjects by ^{1}H NMR spectroscopy, *Metabolomics*, 2006, **2**(3): 105–112.

[97] D Baechle *et al.*, Cathepsin D is present in human eccrine sweat and involved in the postsecretory processing of the antimicrobial peptide DCD-1L, *Journal of Biological Chemistry*, **281**(9): 5406–5415.

[98] SW Brusilow and EH Gordes, Secretion of urea and thiourea in sweat, *Americal Journal of Physiology*, 1965, **209**(6): 1213–12818.

[99] DJ Penn *et al.*, Individual and gender ?ngerprints in human body odour, *Journal of the Royal Society Interface*, 2007, **4**(13): 331–340.

[100] E Marshall, A bruising battle over lung scans, *Science*, 2008, **320**: 600–603.

putting the total cost of a CT-based US lung cancer screening program above $10 billion. Additional costs would be incurred in follow-up testing of false positives.

To evaluate the effectiveness of such a program, the US National Cancer Institute nitiated a 50,000-person trial in 2002. The National Lung Screening Trial[101] is not expected to generate results before 2010, and will likely cost in the region of $200 million. For many other conditions, raising such sums for controlled trials and screening programs will be challenging.

One way to improve the accuracy of biomarker-based predictions is to combine data from multiple biomarkers (multivariate screening). Disease classification using multiple biomarkers is generally much easier and more reliable, in the same way that recognition of a person is much easier from multiple facial features than from a single feature such as an eye.

If the algorithm by which data from multiple biomarkers are combined to arrive at a prediction is complex, additional regulatory clearance may be required prior to marketing of the test.[102] Testing the validity of multivariate assays may therefore require expensive, large-scale clinical trials. However, these costs will be amortized over the total number of users. Thus, widespread adoption of screening and monitoring tests will reduce costs.

As discussed in Chapters 3 and 4, there can be considerable variations in concentrations or activity levels of gene products in different individuals, and in the same individual over time. This consideration complicates the use of quantitative biomarkers. For example, Troponin is a complex of three proteins found in different forms in various muscle cells. Blood-borne cardiac Troponin is often measured by an immunoassay as an indicator of damage to heart muscle arising from a myocardial infarction (MI).[103]

Although elevated blood Troponin has been shown to be a useful indicator of MI, its concentration varies widely among MI candidates reporting chest pain, as shown in figure 7.6 (ibid.). Here, the gray bars indicate measured cardiac troponin concentration in patients with either non-Q-wave myocardial infarction or unstable angina reporting chest pain. The white bars show troponin levels in a control group. Note the greater than ten-fold variation in troponin levels among the MI candidates.

---

[101] http://www.nci.nih.gov/nlst

[102] US Food and Drug Administration Office of *In Vitro* Diagnostic Device Evaluation, *In Vitro* Diagnostic Multivariate Index Assays — draft guidance for industry, clinical laboratories, and FDA staff, June 2006, available at http://tinyurl.com/FDA-guidance-pdf.

[103] EM Antman *et al.*, Cardiac-specific troponin I levels to predict the risk of mortality in patients with acute coronary syndromes, *New England Journal of Medicine*, 1996, **335**: 1342–1349. Figure reproduced with permission.

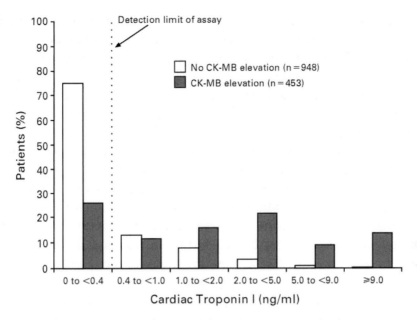

Figure 7.6: Cardiac troponin levels among MI candidates.

The causes of such variations include physiological factors such as circadian rhythms and the effects of age, gender, ethnicity, drugs, diseases, and lifestyle (smoking, alcohol intake, diet, exercise and obesity), as discussed in Chapter 4. Other factors include genetic background (see Chapter 3), patient preparation (e.g. fasting or non-fasting), specimen preparation protocols, and storage conditions.

Patient preparation and sample handling considerations apply to both complex advanced technologies (e.g. mass-spectrometry proteomics, see discussion of the plasma proteome) and also to well-established routinely performed procedures such as histology. For example, a comparison of 200 laboratories worldwide performing the same immunohistochemical assay on a pre-characterized panel of breast cancer samples found that only 37% of the laboratories scored adequately on low-expressing tumors.[104]

Widespread inter- and intra-individual variability makes interpretation of single-sample biomarker data difficult. It would be more useful to take a sequence of readings from the

---

[104] Reliability of immunohistochemical demonstration of oestrogen receptors in routine practice: interlaboratory variance in the sensitivity of detection and evaluation of scoring systems, *Journal of Clinical Pathology*, 2000, **53**: 125–130.

patient and look for trends. This need for repeated biomarker measurements leads to a requirement for monitoring technologies that are minimally invasive, have negligible adverse effects, and are rapid and low-cost. As the preceding review shows, a number of new technologies with these characteristics are emerging.

Microfluidic platforms that measure multiple biomarkers in small quantities of blood, sweat, tear, or saliva are showing great promise.[105] Other technologies, such as microdialysis,[106] offer complementary opportunities for tissue-specific monitoring.

In microdialysis, a needle bearing a small catheter is inserted into the tissue of interest. The needle is removed, leaving the catheter embedded in the tissue for periods of hours to weeks. A permeable membrane in the catheter allows sampling of microliter quantities of size-selected molecules from the tissue interstitial fluid. Microdialysis has its origins in animal neuropharmacology, but can now be used clinically to measure analytes in most human tissues (e.g. skin, heart, brain, breast, bone, and the gastrointestinal tract). Example current applications include continuous glucose monitoring in diabetics and ongoing measurement of drug concentrations within tumors during chemotherapy.

Microdialysis is minimally invasive, widely applicable and fairly low-cost. Samples can be collected/analyzed at intervals as short as a few minutes over extended periods up to a few weeks. Currently, the limiting factor is the availability of assays for a broad range of analytes, but there are no fundamental barriers to the development of such assays and it is likely more will be developed as demand rises.

---

[105] Reviewed in P Yager *et al.*, Microfluidic diagnostic technologies for global public health, *Nature*, 2006, **442**: 412–418.

[106] Reviewed in M Müller, Science,medicine, and the future — Microdialysis, *British Medical Journal*, 2002, **324**: 588–591; CD Anderson, Cutaneous microdialysis: is it worth the sweat? *Journal of Investigative Dermatology*, 2006, **126**: 1207–1209.

CHAPTER 8

Characterizing the Effects of Genetic and Environmental Factors on Cellular Function

The previous chapters showed that variations in genomic sequence and life-long exposures can endow individuals with different susceptibilities to diseases, different mechanisms (e.g. mutations) leading to the same clinical diagnosis, and different responses to treatments. We also reviewed the technologies that allow characterization of an individual's genomic and environmental exposures. This chapter discusses the way in which personal genomes can be combined with life-long health records and multi-parameter molecular diagnostics to provide early detection of disorders, and insights into the underlying mechanisms.

Efficient exploitation of life-long health data requires centralized electronic record-keeping. There is already considerable momentum in the health services industry to implement computerized, multi-user, patient-centered, and comprehensive health records. Once such resources become established, an individual's health history (e.g. past exposures to toxic agents, or responses to specific drugs) can be used to inform future diagnoses and treatment. It will also be possible to mine large-scale patient databases to identify sub-types of disorders and group patients into sub-populations. We will discuss these topics and the ways in which they will improve and personalize healthcare in the next chapter.

The rest of this chapter will assume that health history is taken into account when making diagnostic and prescriptive decisions. We will focus on (1) translating personal genomes into susceptibility predictions; (2) interpreting quantitative biomarker measurements to infer disordered pathways and processes.

Biomedical research and medical practice make different demands on the above aims, as summarized in Table 8.1. This chapter reviews how current biomedical research is addressing these aims. We will discuss the issues relating to the clinical use of these approaches in Chapter 9.

Table 8.1: The differing demands of biomedical research and medical practice on personalized medicine.

Research	Medical practice
Discover sequence variants predisposing to disorders; elucidate mechanism	Find trustworthy research results relevant to patient's individual genomic sequence
Discover biomarkers (and develop related technologies) for early detection	Combine genomic, life-history, and biomarker data into a diagnosis
Develop drugs and other medical intervention strategies	Compare efficacy, adverse-effects, costs, and other aspects of treatment options

How Completely Is the Reference Human Genome Characterized?

To predict the effect(s) of a sequence variant on cellular function, we first need to know the "normal" function(s) of the corresponding DNA region in the reference human genome. In this section, we look at the extent to which we currently understand the cellular role(s) of functional sequences in the reference human genome.

Sequencing of the human genome is *nearly* complete. The 2006 NCBI Build 36 of the human genome covers just under 93% of the reference human genome sequence.[1] Almost all of the unsequenced regions of the genome are heterochromatic. While these regions are critical for chromosomal integrity, cell division, and therefore cellular viability, they are not thought to contribute directly to cellular physiology. However, there are 234 gaps (average length ~70 Kbps) in the available euchromatic DNA sequence, corresponding to about 0.5% of the total genome. Although these gaps are not uniformly distributed, they constitute a small portion of any one chromosome, as summarized in figure 8.1.

Less than 20% of the genome is currently associated with physiologically relevant RNA transcription and related functions. A variety of experimental and computational techniques have been used to annotate the reference genome with features such as exons, introns, transcript start, end and splice sites, repeat sequences, regulatory regions, and the roles of expressed sequences in cellular pathways/processes. The confidence with which DNA sequence features are identified varies considerably across the genome.[2] Some portions — such

---

[1] See http://www.genome.ucsc.edu/goldenPath/stats.html for details.

[2] MR Brent, Steady progress and recent breakthroughs in the accuracy of automated genome annotation, *Nature Reviews Genetics*, 2008, **9**: 62–73; PFR Little, Structure and function of the human genome, *Genome Research*, 2005, **15**: 1759–1766.

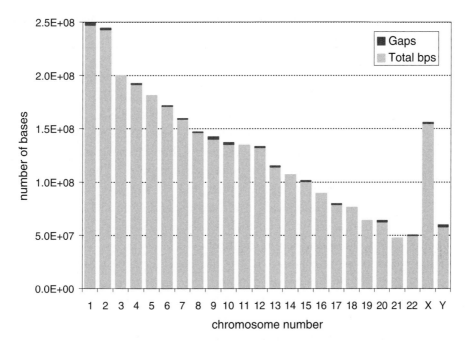

Figure 8.1:   Extent of sequencing gaps per chromosome in the human genome.

Table 8.2:   Key annotated features in the reference human genome.

Known protein-coding genes:	21,388
Number of distinct transcripts:	62,877
Total number of exons:	297,252
RNA genes:	5,732
Pseudogenes (non-functional):	9,899

as those encoding widely studied genes — have been characterized in great detail, but the function and significance of much of the genome remains uncertain. Table 8.2 summarizes the currently known key genomic features of the reference human genome.[3]

Although some of the above numbers seem large, their frequency of occurrence in the genome is surprisingly low. For example, exons occur on average about once in every 10,000

---

[3] February 2009 data downloaded from http://www.ensembl.org/Homo_sapiens/Info/StatsTable.

bases and account for less than 2% of the genome (see Chapter 2). We noted the structural and cell-division roles of chromosomal centromeres and telomeres in Chapter 2. What other functional sequences occupy the human genome?

More than 70% of the human genome is transcribed into non-coding RNAs that are not translated into proteins.[4] Most of these RNAs currently have no known function, but non-coding RNA genes include the functionally important transfer RNA (tRNA) and ribosomal RNA (rRNA) families, as well as a variety of short and long RNAs, many of which appear to regulate gene expression through diverse mechanisms. For example, microRNAs (miRNAs) are typically 20–23 base pairs long and regulate protein synthesis.[5] The mirBase database currently lists 711 native miRNAs.[6] At the other end of the size scale, large intergenic non-coding RNAs (lincRNAs) are typically multi-exonic and span >5 Kbps. More than 1,250 lincRNAs have been found to be active in four mouse embryonic cell types.[7] Many of them appear to be involved in transcriptional regulation.

Because the majority of transcribed RNAs are considered not functionally important, only between 5% and 20% of the human genome is thought to encode physiologically relevant RNA transcription and related functions (Pheasant and Mattick, 2007, Ref. 4). Using this estimate, given a newly sequenced personal genome, it will be necessary to evaluate the potential functional consequences of between 1-in-20 and 1-in-5 of the millions of expected DNA sequence variants. This requirement is an underestimate since a single DNA region may encode multiple transcripts and protein isoforms.

Of course, the vast majority of sequence variants found in any individual genome will not be unique in the human population. After the first several thousand personal genomes have been annotated, the vast majority of functional human DNA variants will

---

[4] M Pheasant and JS Mattick, Raising the estimate of functional human sequences, *Genome Research*, 2007, **17**: 1245–1253. The ENCODE Consortium, Identification and analysis of functional elements in 1% of the human genome by the ENCODE pilot project, *Nature*, 2007, **447**: 799–816; A Petherick, The production line, *Nature*, 2008, **454**: 1043–1045.

[5] A Eulalio, E Huntzinger and E Izaurralde, Getting to the root of miRNA-mediated gene silencing, *Cell*, 2008, **132**: 9–14; D Baek *et al.*, The impact of microRNAs on protein output, *Nature*, 2008, **455**: 64–71; O Hobert, Gene regulation by transcription factors and MicroRNAs, *Science*, 2008, **319**: 1785–1786; EV Makeyev and T Maniatis , Multi-level regulation of gene expression by microRNAs, *Science*, 2008, **319**: 1789–1790; K Hirota *et al.*, Stepwise chromatin remodelling by a cascade of transcription initiation of non-coding RNAs, *Nature*, 2008, **456**: 130–134.

[6] http://microrna.sanger.ac.uk/ — accessed February 2009.

[7] M Guttman *et al.*, Chromatin signature reveals over a thousand highly conserved large non-coding RNAs in mammals, *Nature*, 2009, **458**(7235): 223–227.

likely have been encountered, making the annotation of further personal genomes far simpler.

Less than a quarter of the human proteome has been functionally categorized. The February 2009 release of the UniProtKB/SwissProt database[8] contains manually curated annotations for proteins encoded by 20,331 human genes, with an additional 13,756 isoforms arising from alternative splicing, multiple transcription start sites, or transcription from different promoters (i.e. a total 34,087 distinct proteins). Ultimately, the Human Proteome Initiative intends to populate UniProtKB with descriptions of the function, 3D structure, sub-cellular location, post-translational modifications, variants, and homologies for all human proteins.[8] But at present, the annotations for many proteins are only partial. For example, the structures of only ~15% of the proteins in the database have been experimentally characterized.[8]

A popular way to characterize a protein[9] is through the use of Gene Ontology (GO) categories.[10] GO classifies genes and their products in terms of their molecular function, biological process, and cellular component using a well-defined classification hierarchy. Figure 8.2 shows an example GO annotation hierarchy as a graph.[11] Here the molecule characterized is the T cell receptor alpha chain V region (selected simply because the resulting graph is small enough to be captured within a single page, yet sufficiently detailed to be instructive).

At the top of the graph are the three key categories defined in GO (from left to right: Biological Process, Cellular Component, and Molecular Function). Different border patterns and backgrounds distinguish the three categories in the figure. At the bottom of the figure is the node representing the molecule of interest (TVA2). The remaining nodes represent GO sub-categories at various levels of specificity. As we will see shortly, each of these sub-categories has a unique identifier and associated description (not shown in the figure).

The lines linking the nodes indicate the hierarchical classification relationships. For instance, "membrane" is a sub-category of "cell part". In the case of our example molecule, it is known that TVA2 resides in the plasma membrane (the node directly above TVA2), which is a sub-category of the cellular membrane, which is a sub-category of the cell. In this

---

[8] http://www.expasy.ch/sprot/hpi/

[9] See for example I Friedberg, Automated protein function prediction — the genomic challenge, 2006, *Briefings in Bioinformatics*, 7(3): 225–242.

[10] http://www.geneontology.org/

[11] Graph generated using the GenNav program: http://mor.nlm.nih.gov/perl/gennav.pl.

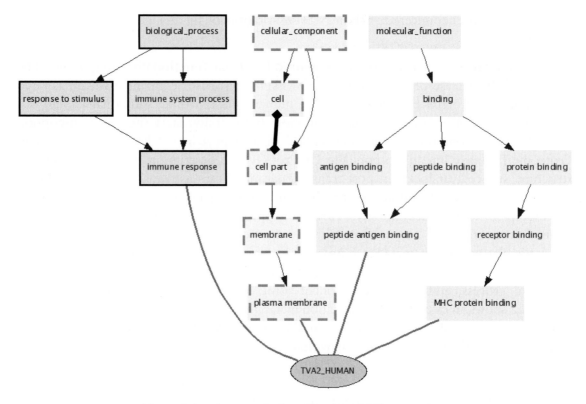

Figure 8.2: An example Gene Ontology (GO) annotation.

way, suitably specific or general categorical descriptors can be assigned to any given molecule depending on the amount of information available.

GO annotations additionally include Evidence Codes (not shown on the graph) which indicate the type and trustworthiness of each assignment.[12] Table 8.3 lists the GO evidence codes in order of descending trustworthiness for each category. Only annotations supported by experimental evidence or a computational prediction with code ISS (see table) are considered high-confidence.[13] These comprise 17.7% of all GO annotations in UniProtKB.

---

[12] Full description available at http://www.geneontology.org/GO.evidence.shtml.

[13] C Yu *et al.*, The development of PIPA: an integrated and automated pipeline for genome-wide protein function annotation, *BMC Bioinformatics*, 2008, **9**: 52.

Table 8.3:   GO evidence codes.

Experimental Evidence Codes

EXP	Inferred from Experiment
IDA	Inferred from Direct Assay
IPI	Inferred from Physical Interaction
IMP	Inferred from Mutant Phenotype
IGI	Inferred from Genetic Interaction
IEP	Inferred from Expression Pattern

Computational Analysis Evidence Codes

ISS	Inferred from Sequence or Structural Similarity
ISO	Inferred from Sequence Orthology
ISA	Inferred from Sequence Alignment
ISM	Inferred from Sequence Model
IGC	Inferred from Genomic Context
RCA	inferred from Reviewed Computational Analysis

Author Statement Evidence Codes

TAS	Traceable Author Statement
NAS	Non-traceable Author Statement

Curator Statement Evidence Codes

IC	Inferred by Curator
ND	No biological Data available

Automatically-assigned Evidence Codes

IEA	Inferred from Electronic Annotation

Obsolete Evidence Codes

NR	Not Recorded

Table 8.4 summarizes the GO annotations for the 42,682 human proteins in the Integr8 database,[14] which amalgamates data from UniProtKB and other sources.[15] The numbers on the left-hand side of the table are the GO identifiers for each category. The two columns on the right-hand side show the numbers and percentages of proteins labeled within the corresponding GO category.

---

[14] http://www.ebi.ac.uk/integr8/, accessed February 2009.

[15] NJ Mulder *et al.*, In Silico characterization of proteins: UniProt, InterPro and Integr8, *Molecular Biotechnology*, 2008, **38**: 165–177.

Table 8.4: GO annotation statistics for human proteins in the Integr8 database.

GO:0003674 molecular function	**21877**	**57.50%**
GO:0003676 nucleic acid binding	4,483	11.70%
GO:0030528 transcription regulator activity	1620	4.20%
GO:0003774 motor activity	220	0.50%
GO:0003824 catalytic activity	7,355	19.30%
GO:0030234 enzyme regulator activity	901	2.30%
GO:0005198 structural molecule activity	963	2.50%
GO:0005215 transporter activity	2,128	5.50%
GO:0004871 signal transducer activity	3936	10.30%
GO:0005488 binding	*14,207*	*37.30%*
GO:0005554 molecular function unknown	*878*	*2.30%*
GO:0008150 biological_process	**18,507**	**48.60%**
GO:0008152 metabolism	10,376	27.20%
GO:0006810 transport	3,543	9.30%
GO:0016265 death	657	1.70%
GO:0006928 cell motility	236	0.60%
GO:0006950 response to stress	1,156	3.00%
GO:0007049 cell cycle	767	2.00%
GO:0007154 cell communication	4,205	11.00%
GO:0007275 development	1,952	5.10%
GO:0007582 physiological process	*15,477*	***40.60%***
GO:0000004 biological process unknown	*735*	*1.90%*
GO:0005575 cellular_component	**16,756**	**44.00%**
GO:0005576 extracellular region	1,394	3.60%
GO:0030312 external encapsulating structure	10	0.00%
GO:0005941 unlocalized protein complex	81	0.20%
GO:0008372 cellular component unknown	823	2.10%
GO:0005623 cell	*15,071*	***39.60%***

Ignoring the categories that are too general to be informative (italicized entries), we note that just over 55% of the entries in the UniProtKB/SwissProt database have been assigned informative Molecular Function and Biological Process sub-categories. As discussed in Chapter 2, ~95% of human genes are alternatively spliced, resulting in an estimated 100,000 distinct proteins. Thus, the current content of the Integr8 database represents ~42% of the human proteome, and has useful functional annotations for less than one quarter (~23%) of the full human proteome. We noted earlier that the more stringent UniProt database currently covers about 34% of the human proteome and has high confidence annotations for only 17.7% of its entries (~6% of the human proteome). Taken together, these observations suggest that between 6% and 23% of the human proteome currently have useful annotations for sequence-variation purposes.[16]

Transcriptional regulatory modules and elements are dispersed throughout the genome. Transcriptional regulatory modules, often referred to as *cis*-regulatory modules (CRMs), contain clusters of transcription factor binding sites and are typically a few hundred to a few thousand nucleotides in length.[17] About one quarter of all transcriptional regulatory modules are thought to be more than 2,500 bps from the nearest transcription start site,[18] making them difficult to locate.

Until recently, it was difficult to identify genomic regions containing transcriptional regulatory modules (enhancers, promoters, repressors, and repressor boundary elements) and transcription factor binding sites. Computational predictions based on multi-species sequence conservation tended to miss regulatory elements/modules responsible for species-specific expression. On the other hand, species-specific computational searches for spatial clusters of binding sites, or shared binding sites among co-expressed genes, tended to generate too many false positives.[19] Meanwhile, the most promising high-throughput experimental technology, the use of Chromatin Immuno Precipitation (ChIP) followed by hybridization of the ChIP-enriched fragments to genomic tiling arrays (ChIP-chip),[20] was proving too difficult, noisy and expensive to allow comprehensive genomic annotation.[21]

The advent of low-cost high-throughput short-read sequencing technologies has dramatically changed this situation. Instead of hybridizing to a genomic tiling array, ChIP

---

[16] Consistent with the above conclusion, a recent study scoring annotations according to six criteria found 5,286 out of 33,410 predicted genes (15.8% of genes, or ~5.3% of the estimated full proteome) have high annotation levels (a score of 7.5 or higher out of 10). D Kemmer *et al.*, Gene characterization index: assessing the depth of gene annotation, *PLoS One*, 2008, **1**: e1440.

[17] EH Davidson, *The Regulatory Genome*, Academic Press, 2006.

[18] The ENCODE Project consortium; see Ref. 4.

[19] For a review of the issues, see WW Wasserman and A Sandelin, Applied bioinformatics for the identification of regulatory elements, *Nature Reviews Genetics*, 2004, **5**: 276–287.

[20] MJ Buck and JD Lieb, ChIP-chip: considerations for the design, analysis, and application of genome-wide chromatin immunoprecipitation experiments, *Genomics*, 2004, **83**: 349–360.

[21] Several array chips are required to tile the entire human genome, and cross-hybridization tends to generate considerable background noise, especially if the tiling array tags are short and not carefully selected.

fragments are now directly sequenced at one end and mapped onto the reference genome. The number of sequenced ChIP fragments at any given genomic location indicates the level of ChIP enrichment at that address.

This approach (labeled ChIP-seq) avoids cross-hybridization noise and the high cost of multi-chip whole-genome tiling arrays. Moreover, size selection of ChIP fragments to 100–150 bp lengths has allowed delineation of transcription factor binding sites to within 100 bp windows.[22] These regions are small enough that simply scanning for known binding motifs, or searching for statistically over-represented motifs among multiple marked windows, unambiguously identifies transcription factor binding sites.[22]

A limiting factor in the use of Chromatin IP has been the availability of suitable antibodies for diverse transcription factors. However, antibodies which bind specifically to various types of modified histones are providing a solution. As noted in Chapters 2 and 4, chromatin state is regulated through various chemical modifications to histone proteins. It has recently become clear that particular combinations of histone modifications can be used to identify proximal promoters, distal enhancers, and active or repressed transcriptional states in transcribed regions.[23]

Performing ChIP-seq using modification-specific histone antibodies is rapidly delivering genomic maps of regulatory sequences. DNA regions delineated in this way are sufficiently well defined to allow high-confidence computational identification of transcription factor binding sites within them.[24]

Thus, it is increasingly possible to confidently identify promoters and enhancers, and predict the binding sites of transcription factors, even for factors lacking suitable antibodies.

---

[22] DA Johnson *et al.*, Genome-wide mapping of *in vivo* protein-DNA interactions, *Science*, 2007, **316**: 1497–1502.

[23] ND Heintzman *et al.*, Distinct and predictive chromatin signatures of transcriptional promoters and enhancers in the human genome, *Nature Genetics*, 2007, **39**(3): 311–318; TS Mikkelsen *et al.*, Genome-wide maps of chromatin state in pluripotent and lineage-committed cells, *Nature*, 2007, **448**: 553–560; A Barksi *et al.*, High-resolution profiling of histone methylations in the human genome, *Cell*, 2007, **129**: 823–837; A Visel *et al.*, ChIP-seq accurately predicts tissue-specific activity of enhancers, *Nature*, 2009, **457**: 854–859.

[24] R Jothi *et al.*, Genome-wide identification of *in vivo* protein–DNA binding sites from ChIP-seq data, *Nucleic Acids Research*, 2008, **36**(16): 5221–5231; T Whitington, AC Perkins and TL Bailey, High throughput chromatin information enables accurate tissue-specific prediction of transcription factor binding sites, *Nucleic Acids Research*, 2009, **37**(1): 14–25.

Moreover, because of its low cost, ChIP-seq can be used to identify the cell types and conditions under which specific *cis*-regulatory elements/modules are active.[25]

We noted earlier that a large fraction of the human genome expresses non-coding RNAs. ChIP-seq techniques can equally be used to characterize the regulation of activity of these sequences also. In short, although current annotation of human regulatory sequences is sparse,[26] we can expect this picture to change very rapidly over the next few years.

Functional Annotation of Genomic Sequences

The availability of a high-quality reference human genome allows fast and low-cost assembly of personal genomes using high-throughput technologies. Comparative genomic hybridization and paired-end sequencing (see Chapter 6) can be used to identify any large-scale (structural) differences between a newly sequenced genome and the reference genome.

We can also identify previously characterized functional elements (e.g. regulatory sequences, exons, introns, etc.) in newly assembled personal genome using the reference sequence.

Where sequence variations affect annotated functional elements, we can evaluate the implications in two steps. First, we must determine the impact of each sequence variant on the molecules directly affected. Second, we must predict the ways in which changes in abundance or biochemical properties of any affected molecules will impact cellular and organ physiology.

In the simplest case, a variant sequence may match an existing entry in mutation databases such as OMIM (http://www.ncbi.nlm.nih.gov/Omim/) and HGMD (http://www. hgmd.cf.ac.uk/). As of February 2009, HGMD contains data on 85,558 distinct mutations, as summarized in Table 8.5.

We note two features in Table 8.5. Firstly, as discussed in Chapter 3, any individual's genome is likely to differ from the reference genome by as much as one in every

---

[25] See for example G Robertson *et al.*, Genome-wide profiles of STAT1 DNA association using chromatin immunoprecipitation and massively parallel sequencing, *Nature Methods*, 2007, **4**(8): 651–657; R Nielsen *et al.*, Genome-wide profiling of PPARγ:RXR and RNA polymerase II occupancy reveals temporal activation of distinct metabolic pathways and changes in RXR dimer composition during adipogenesis, *Genes and Development*, 2008, **22**: 2953–2967.

[26] For example databases of human *cis*-regulatory elements and modules (with different evidence requirements), see http://enhancer.lbl.gov/; http://rulai.cshl.edu/TRED; http://www.cisred.org/human9/; and http://www.pazar.info.

Table 8.5: Summary of mutation data in the Human Genome Mutation Database.

Missense/nonsense	Single base-pair substitutions in coding regions are reported as a triplet change.	48,343
Splicing	Positions of mutations with consequences for mRNA splicing are reported.	8,219
Regulatory	Locations of substitutions causing regulatory abnormalities are specified.	1,400
Small deletions	Micro-deletions of 20 bp or less.	13,628
Small insertions	Micro-insertions of 20 bp or less.	5,567
Small indels	Micro-indels of 20 bp or less.	1,244
Gross deletions	Reported in narrative form because of wide variations in reporting style in the literature.	5,158
Gross insertions	Reported in narrative form because of wide variations in reporting style in the literature.	1,003
Complex rearrangements	Reported in narrative form because of wide variations in reporting style in the literature.	736
Repeat variations	Reported in narrative form because of wide variations in reporting style in the literature.	260

500 nucleotides. In the preceding section in this chapter, we saw that up to 20% of these variations may have functional consequences. Thus, a newly sequenced personal genome may have in excess of one million sequence variations that must be evaluated for potentially functional consequences.

We see that the total number of currently known functional mutations, as summarized in HGMD, is about an order of magnitude smaller than the number of variants affecting functional sequences in a personal genome. Even if a large majority of these variants turn out to be functionally neutral (benign), *de novo* evaluation of the potential effects of many sequence variations will be necessary.

The second observation arising from the above table is that a large number of the HGMD entries are stored in the form of narrative text that will be difficult to exploit automatically by computational methods (similar observations hold for aspects of OMIM). Thus, on the one hand, the numbers of sequence variants that will need to be evaluated per personal genome are so large that at least initial genome-wide analysis must be automated. On the other hand, existing databases cover a small proportion of the total number of sequence variants that will need to be analyzed, and even this information is not stored in a format that lends itself well to automation.

Below, we review emerging computational and technological approaches that aim to determine the function of all hitherto uncharacterized expressed sequences, and predict the functional effects of previously uncharacterized sequence variations.

Predicting the Effects of Sequence Variants

Predicting the functions and interactions of untranslated RNAs. As of February 2009, the abundances, types, interactions, and cellular roles of most untranslated RNAs remain the subject of ongoing research.[27] However, for some better characterized non-coding transcripts, such as micro RNAs, there are already several databases of known human transcripts and their targets.[28] Moreover, a multitude of computational tools have been developed to predict the targets of a given miRNA based on features such as thermodynamic stability, target site accessibility, and evidence of evolutionary conservation of the miRNA's structure.[29] Similar resources, and improved prediction accuracies can be expected for other types of untranslated RNAs as we learn more about them.

Predicting mRNA stability changes. Human mRNA half-lives range from about 20 minutes to four days. Mutations that result in premature translation termination sites can destabilize mRNAs by ten-fold or more.[30] For wild-type mRNAs, the rate of degradation in the cytoplasm is regulated by binding motifs on the mRNA (*cis*-elements), and by the proteins that bind these motifs (trans-acting factors).[31] Thus, regulation of mRNA stability is in some ways similar to regulation of gene transcription. We will discuss the effects of mutations in *cis*-regulatory elements and trans-acting proteins below.

Regulatory proteins can increase as well as decrease mRNA stability. As we noted earlier, a variety of regulatory RNAs can additionally mediate condition-specific degradation

---

[27] See for example P Carninci, The long and short of RNAs, *Nature*, 2009, **457**: 974–975; The Affymetrix/Cold Spring Harbor Laboratory ENCODE Transcriptome Project, Post-transcriptional processing generates a diversity of 5′-modified long and short RNAs, *Nature*, 2009, **457**: 1028–1032.

[28] For example http://microrna.sanger.ac.uk/index.shtml and http://diana.cslab.ece.ntua.gr/tarbase.

[29] See for example http://bibiserv.techfak.uni-bielefeld.de/rnahybrid/submission.html; http://www.microrna.org/microrna/getGeneForm.do; http://www.targetscan.org/. For a general review see IM Meyer, Predicting novel RNA–RNA interactions, *Current Opinion in Structural Biology*, 2008, **18**: 387–393.

[30] J Ross, mRNA stability in mammalian cells, *Microbiological Reviews*, 1995, **95**(3): 423–450.

[31] See for example NH Ing, Steroid Hormones regulate gene expression post transcriptionally by altering the stabilities of messenger RNAs, *Biology of Reproduction*, 2005, **72**: 1290–1296.

of mRNAs or block their translation. For miRNAs, software packages that predict binding targets[29] can be used to check if target binding is affected by a variant. More generally, because miRNAs are short and have a relatively simple structure, mRNA folding algorithms can predict the effects of sequence variants on their structure with high accuracy.[32]

Predicting the effects of coding sequence variations. It has been estimated that a typical individual will be heterozygous for about 24,000–40,000 non-synonymous (amino acid altering) substitutions.[33] Because of the large numbers involved, until recently genome-wide experimental evaluation of SNP effects on protein structure and function has not been possible. Instead, a variety of computational approaches are used to predict candidate deleterious coding SNPs. Currently available computational methods typically exploit two types of information:[34]

(1) Evolutionary conservation: coding nucleotides that are highly conserved within protein families or across species are likely to be critical to protein function.
(2) Structural significance: Amino acids buried inside a protein structure are more likely to be essential to its structural integrity. On the other hand, externally facing residues are more likely to be involved in interactions.

     An alternative to predicting interaction sites from structural information is to search public databases (such as SwissProt) for known binding sites. If the structure of a protein of interest is not available, the structure of a homologous protein can be used instead. However, prediction accuracy drops with lower similarity.

The two approaches use complementary sources of evidence and provide complementary clues to the potential effects of a sequence variant. For example, sequence-based methods may detect alternative splicing effects which would not be apparent with structure-only analysis. On the other hand, structural methods can reveal important topological surface

---

[32] BA Shapiro *et al.*, Bridging the gap in RNA structure prediction, *Current Opinion in Structural Biology*, 2007, **17**: 157–165; AD George and SA Tenenbaum, Informatic resources for identifying and annotating structural RNA motifs, *Molecular Biotechnology*, 2009, **41**: 180–193.
[33] M Cargill *et al.*, Characterization of single-nucleotide polymorphisms in coding regions of human genes, *Nature Genetics*, 1999, **22**: 231–238.
[34] PC Ng and S Henikoff, Predicting the effects of amino acid substitutions on protein function, *Annual Review of Genomics and Human Genetics*, 2006, **7**: 61–80.

features (e.g. accessible protein–protein interaction domains). Current computational methods detect about half the known deleterious SNPs, and overlap in only about one in 25 predictions.[35]

In addition to sequence and structure based predictions, two other approaches have proved fruitful. First, mutations that affect the function (but not necessarily abundance) of transcription factors can be detected from expression changes in affected target genes.[36] Second, text-mining and database searching can provide supporting or complementary data from the literature (e.g. post-translational modification sites, cellular localization signals, and transmembrane components[37]).

The precision of computational predictions of SNP effects depends critically on the amount of sequence-feature, structural, and functional data available on the wild-type protein of interest and its homologs.[38] A particular challenge at present is that most of the training data currently available for SNP-effect prediction relates to high-penetrance, rare Mendelian alleles. Such SNPs are strongly selected against (hence their rarity), and the corresponding wild-type nucleotide exhibits considerable evolutionary conservation.[39] In contrast, SNPs that contribute to multigenic disorders appear not to be individually selected against, and their wild-type counterparts do not seem to be evolutionary conserved.[40]

Thus, predictions of the effects of sequence variants within complex-disease loci cannot rely on evolutionary conservation information. This may explain why current SNP-effect prediction tools do not predict deleterious effects for many SNPs documented in dbSNP (http://www.ncbi.nlm.nih.gov/projects/SNP/) and SwissProt (http://ca.expasy.org/sprot/).

---

[35] DF Burke *et al.*, Genome bioinformatic analysis of nonsynonymous SNPs, *BMC Bioinformatics*, 2007, **8**: 301.

[36] NJ Hudson, A Reverter and BP Dalrymple, A differential wiring analysis of expression data correctly identifies the gene containing the causal mutation, *PLoS Computational Biology*, 2009, **5**(5): e1000382.

[37] See for example http://SNPeffect.vib.be, described in J Reumers *et al.*, SNPeffect v2.0: a new step in investigating the molecular phenotypic effects of human non-synonymous SNPs, *Bioinformatics*, 2006, **22**(17): 2183–2185.

[38] D Tchernitchko, M Goossens and H Wajcman, In silico prediction of the deleterious effect of a mutation: proceed with caution in clinical genetics, *Clinical Chemistry*, 2004, **50**(11): 1974–1978.

[39] Interestingly, indels appear to be strongly selected against in some protein classes and not others. So prediction of their effects will require more nuanced analyses. See N de la Chaux, PW Messer and PF Arndt, DNA indels in coding regions reveal selective constraints on protein evolution in the human lineage, *BMC Evolutionary Biology*, 2007, 7: 191.

[40] PD Thomas and A Kejariwal, Coding single-nucleotide polymorphisms associated with complex vs. Mendelian disease: evolutionary evidence for differences in molecular effects, *PNAS*, 2004, **101**(43): 15398–15403.

Finally, the effects of SNPs on protein stability can be predicted with accuracy exceeding 90% of the theoretical optimum, if the wild-type protein structure is available.[41] Remarkably, recently developed methods that use only sequence information still manage to predict the direction of stability change with greater than 75% accuracy, and the magnitude of the change with greater than 60% correlation.[42]

Predicting the functions and interactions of novel proteins. We noted earlier that less than one quarter of all human proteins have been characterized in terms of their biochemical function, the cellular pathways and processes they take part in, and their specific molecular interaction partners.

A number of recent technological developments have increased the throughput and lowered the costs associated with experimental protein structure determination.[43] Using limited experimental data to constrain computational structure predictions is also lowering prediction costs and improving the predictive power of *de novo* structure prediction algorithms.[44] Accordingly, a number of large-scale structure determination projects are underway.[45] High costs limit this approach to candidate proteins of special interest. Nonetheless, rapid progress may be expected. For example, the multi-national Structural Genomics Consortium[46] aims to determine the structure of about 1,000 proteins by 2011.

When possible, genome-wide protein annotation is performed by matching the sequence or secondary structure of the novel protein to those of existing protein families.[47] For example, the PIPA protein annotation pipeline generates GO annotations from genomic sequences by searching against 16 databases of protein sequence features.[48] Feature-based

---

[41] M Masso and II Vaisman, Accurate prediction of stability changes in protein mutants by combining machine learning with structure based computational mutagenesis, *Bioinformatics*, 2008, **24**(18): 2002–2009.

[42] E Capriotti *et al.*, I-Mutant2.0: predicting stability changes upon mutation from the protein sequence or structure, *Nucleic Acids Research*, 2005, Web Server Issue, **33**: W306–W310; L-T Huang *et al.*, iPTREE-STAB: interpretable decision tree based method for predicting protein stability changes upon mutations, *Bioinformatics*, 2007, **23**(10): 1292–1293.

[43] See for example http://www.fluidigm.com/products/topaz-main.html.

[44] R Das and D Baker, Macromolecular modeling with Rosetta, *Annual Review of Biochemistry*, 2008, **77**: 363–382.

[45] For examples see http://www.nigms.nih.gov/Initiatives/PSI/, and http://www.thesgc.com.

[46] http://www.thesgc.com/, see documentation under the "About SGC" tab.

[47] Reviewed in I Friedberg, Automated protein function prediction — the genomic challenge, *Briefings in Bioinformatics*, 2006, 7(3): 225–242.

[48] C Yu *et al.*, The development of PIPA: an integrated and automated pipeline for genome-wide protein function annotation, *BMC Bioinformatics*, 2008, **9**: 52.

protein function prediction can be applied to entire human genomes using relatively modest computing power. Moreover, the likelihood of finding good feature matches increases as more proteins are structurally and biochemically characterized.

A key aim of protein annotation is to identify the cellular processes and pathways that the protein takes part in. Accordingly, a wide variety of experimental and computational methods that predict protein–protein interactions have been developed. The Unified Human Interactome database (UniHI), which aggregates data from 13 sources of experimental and computational protein interaction predictions, currently lists 200,473 interactions among 22,307 unique proteins.[49] However, only 19% of the entries are found in two or more data sources,[50] suggesting potentially high false-positive rates. Consistent with this observation, an analysis of data from eight experimental and computational sources available in 2005 found that only 10% of the data were present in two or more data sources.[51]

To minimize false positives, a number of computational tools rank confidence levels in predicted protein interactions by comparing features of novel predictions against features of validated interactions.[52] Such confidence statistics can be combined to estimate and rank the overall likelihood that a novel protein interacts with various cellular pathways.

It should be noted that the lack of agreement between existing datasets may not be entirely due to false positives. The current set of known proteins and their interactions represents a small fraction of the total expected. Thus, some of the lack of overlap between datasets may be due to each study sampling a different region of the full protein interaction network. As more proteins are characterized, we can expect greater concordance between independently collected datasets.

Predicting the effects of regulatory sequence variations. SNPs in transcriptional regulatory regions may interfere with the corresponding gene's *cis*-regulatory logic. About one third of SNPs falling in promoter regions are thought to significantly affect the expression

---

[49] February 2009 data from http://theoderich.fb3.mdc-berlin.de:8080/unihi/home.

[50] G Chaurasia *et al.*, UniHI 4: new tools for query, analysis and visualization of the human protein–protein interactome, *Nucleic Acids Research*, 2009, Database issue, **37**: D657–D660.

[51] ME Futschik *et al.*, Comparison of human protein–protein interaction maps, *Bioinformatics*, 2007, **23**(5): 605–611.

[52] D Li *et al.*, PRINCESS, a protein interaction confidence evaluation system with multiple data sources, *Molecular and Cellular Proteomics*, 2008, 7(6): 1043–1052; MD McDowall, MS Scott and GJ Barton, PIPs: human protein–protein interaction prediction database, *Nucleic Acids Research*, 2009, Database issue, **37**: D651–D656.

pattern of the associated gene.[53] DNA sequence features at and around the SNP location (e.g. repeat content) can be used to predict whether a SNP in these regions will have a functional effect.[54] For example, the PromoLign software (http://polly.wustl.edu/promolign/main.html) can be used to rank the significance of SNPs in putative regulatory regions according to whether a SNP occurs within (i) a known or predicted transcription factor binding site, (ii) an evolutionary conserved region, and (iii) a region with homologs in other promoters. SNPs located *within* repeat sequences are discounted[55] (as opposed to the regions flanking SNPs, which typically have a higher repeat content[54,56]).

The frequency of observed nucleotides at specific locations within transcription factor binding sites reflects the binding affinity of the transcription factor for each nucleotide.[57] When a SNP falls within a known or putative transcription factor binding site, it causes a small change in the binding affinity of the site. In itself, the estimated change in binding affinity is not large enough to allow robust predictions of SNP effects.[58] However, applied to SNPs in evolutionary conserved sequences, this approach appears useful for predicting functional SNPs.[54,58]

If genome-wide expression data is available, clusters of genes with highly correlated expression patterns are often co-regulated and enriched for particular cellular processes. Tests of SNP effects performed on co-expressed gene clusters typically have greater predictive power.[59] In a similar vein, if a list of transcription factors known to be active in the cell and condition of interest is available, only SNPs occurring in the predicted binding sites of the active factors need be considered.[58]

---

[53] B Hoogendoorn *et al.*, Functional analysis of human promoter polymorphisms, *Human Molecular Genetics*, 2003, **12**(18): 2249–2254.

[54] The discriminatory potential of 23 features was analyzed by SB Montgomery *et al.*, A survey of genomic properties for the detection of regulatory polymorphisms, *PLoS Computational Biology*, 2007, **3**(6): e106.

[55] T Zhao *et al.*, PromoLign: a database for upstream region analysis and SNPs, *Human Mutation*, 2004, **23**: 534–539.

[56] This effect may be due to the widespread role of Transposable Elements (see Chapter 4) in sculpting *cis*-regulatory regions. C Feschotte, Transposable elements and the evolution of regulatory networks, *Nature Reviews Genetics*, 2008, **9**: 397–405.

[57] Reviewed in GD Stormo, DNA binding sites: representation and discovery, *Bioinformatics*, 2000, **16**(1): 16–23.

[58] MA Andersen *et al.*, In silico detection of sequence variations modifying transcriptional regulation, *PLoS Computational Biology*, 2008, **4**: 1.

[59] L Michal, O Mizrahi-Man and Y Pilpel, Functional characterization of variations on regulatory motifs, *PLoS Genetics*, 2008, **4**(3): e1000018; S-I Lee *et al.*, Identifying regulatory mechanisms using individual variation reveals key role for chromatin modification, *Proceedings of the National Academy of Sciences of USA*, 2006, **103**(38): 14062–14067.

Predicting the effects of functional mutations on cellular pathways. In Chapter 5, we reviewed some of the ways in which protein–protein and protein–DNA interactions can lead to complex behaviors. We also presented examples of how changes in component characteristics (e.g. half-lives, or binding affinities) can affect these behaviors. Here, we will briefly review the process by which one might assess changes in cellular behavior given a predicted functional mutation.

The exact procedure for mapping a functional mutation to changes in cellular function will vary depending on the amount of information available about the mutated gene. For poorly characterized genes, we may first need to predict potential functions and interactions using the resources discussed earlier in this chapter. On the other hand, if the affected molecule is well known, its roles in various cellular processes are likely to be documented in existing pathway databases.[60]

As of July 2009, the community-curated Reactome database (http://www.reactome.org/) catalogues 990 pathways involving 4181 proteins and 3335 reactions. Reactome enforces a hierarchy of curation stringency, ranging from "observed in other organisms/cell types" to "characterized *in vivo* in cells of interest". The above figures are for the *total* number of pathway entries.

In terms of high-confidence in-human interactions, the PathwayInteractionDatabase (PID, http://pid.nci.nih.gov/), which includes entries from the 2007 release of Reactome, has a total of 414 curated human pathways (accessed July 2009). In a similar vein, the KEGG database (http://www.genome.ad.jp/kegg/), which is manually derived from the published literature, lists just over 350 pathways associated with ~30 distinct cellular processes (accessed July 2009). Given the pace of developments, most human genes are likely to be documented at this confidence and detail in the near future.

The assignment of genes to pathways can be aided by the availability of additional data. For example, if in addition to sequence data genome-wide expression data is also available, then clusters of co-expressed genes can be associated with particular sequence variants. Often, some of the genes within each co-expression cluster are well known and can be mapped to pathways with high confidence.[61] Cellular pathways discovered in this way can then be considered "candidate affected pathways" for uncharacterized genes in the cluster.[62]

---

[60] A comprehensive directory of all pathway databases is maintained at http://www.pathguide.org. In February 2009, PathGuide listed a total of 112 pathway databases for *Homo sapiens*.

[61] For example, using the Gene Set Enrichment Analysis (GSEA) method: http://www.broad.mit.edu/gsea, A Subramanian *et al.*, Gene set enrichment analysis: a knowledge-based approach for interpreting genome-wide expression profiles, *Proceedings of the National Academy of Sciences of USA*, 2005, **102**(43): 15545–15550.

[62] A Ghazalpour *et al.*, Integrating genetic and network analysis to characterize genes related to mouse weight, *PLoS Genetics*, 2006, **2**(8): e130.

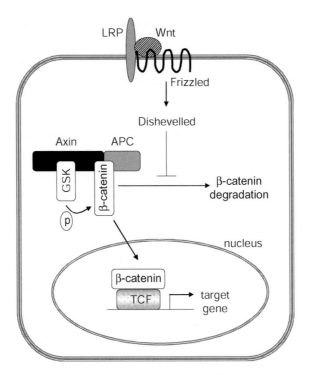

Figure 8.3:   A simplified view of the canonical Wnt signaling pathway.

Thus, mutations in both well-characterized and novel gene products can be mapped to candidate affected cellular processes.

Once a functional mutation has been mapped to specific pathways, an initial assessment of its effect on impacted pathways can usually be made directly from the pathway diagram. For example, the figure 8.3 shows a simplified view of the canonical Wnt signaling pathway. The arrows indicate activation or transport, and the line ending in a bar indicates repression. In the absence of Wnt signaling, Axin-APC complexes form a scaffold which brings together β-catenin and its kinase complex (labeled GSK). Phosphorylated β-catenin is then unbiquiti-nated and degraded. When Wnt ligands bind LRP-Frizzled dimers, they inhibit β-catenin degradation via Dishevelled. β-catenin then accumulates in the nucleus, where it pairs with the transcription factor TCF to activate target genes.

For the highly simplified pathway diagram of figure 8.3, visual inspection is sufficient to deduce that any mutation that disrupts the degradation of β-catenin (e.g. loss of the phos-phorylation site on the β-catenin molecule, or mutations in Axin or APC that disrupt the scaffold or its ability to phosphorylate β-catenin), will lead to accumulation of β-catenin in the nucleus and therefore Wnt-independent activation of Wnt target genes. Observations in colorectal tumors support this hypothesis. Wnt-independent activation of Wnt target genes underlies ~90% of colorectal cancers, and nearly 70% of colorectal cancers have mutations in APC that disrupt β-catenin phosphorylation.[63]

As we saw in Chapters 2 and 5, cellular pathways involve large numbers of gene products, interactions, and feedback loops. Consequently, real pathway diagrams are not as easy to analyze as the simplified example given above (figure 8.3). Moreover, the large numbers of sequence variants that will have to be analyzed in personal genomes will require automated screening.

The convergence of a number of recent developments is making automated analysis of mutation effects in pathways possible. Firstly, new graphical notations being used to draw pathway diagrams now specify each interaction in precise, unambiguous terms.[64]

Secondly, these graphical pathway representations can be translated into machine readable textual descriptions. Indeed, all three pathway databases mentioned earlier support a number of model description languages.[65]

Thirdly, since the graphical notation standards unambiguously define a set of logical relationships between molecular species, pathway diagrams can be translated into formal logic descriptions automatically.[66] Such descriptions can then be subjected to formal logic analyses, for example to explore the effect of a mutation. While these methods are currently only at the "proof of concept" stage, they hold great promise for automated analysis of the effects of the large numbers of mutations that we may need to evaluate in order to characterize personal genomes.

Because of their many feedback loops, cellular pathways can have complex and unintuitive behaviors (see Chapter 5). Logical models can predict qualitative state changes in perturbed

---

[63] H Suzuki *et al.*, Epigenetic inactivation of SFRP genes allows constitutive WNT signaling in colorectal cancer, *Nature Genetics*, 2004, **36**(4): 417–422. See also MM Taketo, Shutting down Wnt signal–activated cancer, ibid.: 320–322.

[64] See for example the Systems Biology Graphical Notation: http://sbgn.org, and the Molecular Interaction Map format: http://discover.nci.nih.gov/mim.

[65] KEGG has its own XML-based description language called KGML (http://www.genome.jp/kegg/xml/). Other common description standards are: SBML (http://sbml.org/), BioPax (http://www.biopax.org/), and CellML (http://www.cellml.org/). KGML can be translated to SBML and other formats.

[66] For an example see J Feret *et al.*, Internal coarse-graining of molecular systems, *Proceedings of the National Academy of the USA*, 2009, **106**(16): 6453–6458.

cellular processes, but they cannot predict the quantitative and kinetic effects of mutations.[67] For example, in the preceding example, it turns out that mutations that disrupt β-catenin phosphorylation only partially activate Wnt target genes independent of Wnt signaling. Indeed, ~90% of colorectal tumors include an additional mutation that results in improved binding of Wnt ligands to Frizzled-LRP receptor dimers.[63]

In general, mutations that change the rate of a process but not its steady-state can have adverse effects — such as toxic build-up — that are not predicted by logical models. One way to capture such effects is to build quantitative and kinetic models of cellular processes and physiology.

Development of mechanistic, biochemical models of cellular physiology can be difficult and time-consuming because of the need to collect biochemical reaction rates and to cross-validate the model's behavior in a range of conditions.[67] But the increasing uptake of integrative systems biology and the need for quantitative predictive models are driving rapid developments in the field.[68] For example, as of February 2009, the Biomodels database (http://www.ebi.ac.uk/biomodels-main/) has over 200 expert-curated biochemical models of cellular processes, including 62 models of about 40 human cellular processes.

Several large-scale projects are currently underway to develop integrative models of human physiology. Prominent among the academic efforts are the well-established Physiome project[69] and the more recent Virtual Physiological Human project,[70] for which the European Union has earmarked funding of 72 million euros.[71] Both projects are developing integrative multi-organ physiological models at the molecular, cellular, organ and organism levels.[72]

For biomedical modeling purposes, coarse-grained empirical models of dynamic interactions among key clinical variables are proving highly successful. This is in part because of the large amounts of clinical data available for model development and validation. The Archimedes system developed at Kaiser Permanente (http://archimedesmodel.com/) offers a good example.

---

[67] Discussed in H Bolouri, *Computational Modeling of Gene Regulatory Networks — A Primer*, Imperial College Press, 2008.

[68] See the systems biology Special Issue of *Science* magazine, 1 March 2002, **295**(5560).

[69] http://www.physiome.org.nz/

[70] http://www.vph-noe.eu/

[71] G Clapworthy *et al.*, Special issue editorial, *Philosophical Transactions of the Royal Society of London A*, 2008, **366**: 2975–2978.

[72] P Kohl *et al.*, The virtual physiological human: tools and applications II, *Philosophical Transactions of the Royal Society of London A*, 2009, **367**: 2121–2123.

Archimedes is an integrative, dynamic model of human physiology built from a large set of empirical relationships among clinically measurable variables such as heart rate, stroke volume, cardiac output, peripheral resistance and blood pressure.[73] It currently offers models of a dozen disorders, including diabetes, various cardiac diseases, asthma, and some cancers. It was originally developed for type 1 and type 2 diabetes, and in this regard it has been tested extensively against clinical trial data and been found to be highly accurate in its predictions.[74]

Archimedes-like physiological models for a range of other medical applications are currently being developed by a number of commercial ventures.[75] In particular, Entelos (http://www.entelos.com/), which has a portfolio of physiological models,[76] has announced a model-based personal health analysis service (http://mydigitalhealth.com/).

From observed symptoms to candidate genes. The methods discussed above attempt to predict the effects of DNA sequence variants on the function and behavior of cellular pathways. A complementary approach is to start with disease symptoms (phenotypes) and attempt to predict correlated DNA sequence variants. In recent years, there have been several exciting developments along these lines. A few examples are reviewed below to provide a flavor of this rapidly developing field.

The OMIM database identifies some mutations as having strong evidence of causal links to named disorders.[77] Goh *et al.*[78] used these entries to create mappings from genes to associated diseases and vice versa. Figure 1.4 in Chapter 1 showing some of the disease-gene associations for Alzheimer's was built using this methodology. As Goh *et al.* point out, the gene-disease associations can be viewed in two complementary ways. The "disease view" indicates the number of causal genes shared between any two diseases, while the "gene view" shows shared diseases associated with given genes.

Figure 8.4 shows a small portion of the disease and gene views of OMIM data.[79] The node labels indicate the disorders (left panel) and genes (right panel) characterized.

---

[73] L Schlessinger and DM Eddy, Archimedes: a new model for simulating health care systems — the mathematical formulation, *Journal of Biomedical Informatics*, 2002, **35**: 37–50.

[74] DM Eddy and L Schlessinger, Archimedes: a trial-validated model of diabetes, *Diabetes Care*, 2003, **26**: 3093–3101; Validation of the Archimedes diabetes model, *Diabetes Care*, 2003, **26**: 3102–3110.

[75] K Thiel, Systems biology, incorporated? *Nature Biotechnology*, 2006, **24**(9): 1055–1057.

[76] See for example Y Zheng *et al.*, The virtual NOD mouse applying predictive biosimulation to research in Type 1 Diabetes, *Annals of the New York Academy of Sciences*, **1103**: 45–62. See also http://realab.entelos.com.

[77] The class 3 tagged entries in the OMIM Morbid Map, http://www.ncbi.nlm.nih.gov/Omim/getmorbid.cgi.

[78] K-I Goh *et al.*, The human disease network, *Proceedings of the National Academy of Sciences of USA*, 2007, **104**(21): 8685–8690.

[79] Figure reproduced from Goh *et al.*, Ref. 78. Copyright (2007) National Academy of Sciences, U.S.A.

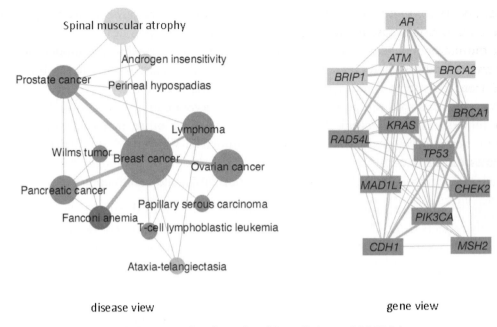

disease view   gene view

Figure 8.4:   Example "disease" and "gene" views of OMIM data.

Node sizes in the disease view indicate the number of genes associated with that disorder. The thickness of the edges connecting the nodes is proportional to the amount of overlap between the nodes (no connection indicates zero overlap). Note the extensive overlap in both the genes shared between diseases and diseases shared by genes.

The observation that many diseases share causal genes, and many genes contribute to shared diseases can be used to select candidate disease genes through "guilt-by-association" approaches. The underlying assumption is that genes that contribute to the same disorder must share some characteristics. Likewise, diseases that share causal genes are hypothesized to share dysregulated cellular processes.

In a generalization of this concept, the candidate gene prediction tool Endeavour[80] allows the user to specify multiple gene/protein features as phenotype predictors. In the calibration phase, user-specified gene features such as GO annotation, protein–protein interactions, co-expression and joint pathway membership are trained to be predictors of known

---

[80] S Aerts *et al.*, Gene prioritization through genomic data fusion, *Nature Biotechnology*, 2006, **24**(5): 537–544.

gene-disease associations. As a result of this training/calibration process, each gene feature can be used to make noisy predictions of gene-disease associations.

In the prediction phase, for each disorder a large number of potentially contributing genes are evaluated and ranked in order of likelihood by each feature-based predictor. The results from all predictors are then merged into an overall ranking of candidate genes. Preliminary tests suggest Endeavour predictions of candidate genes are within ~10% of the optimum classifier.

In an analysis of 1.5 million patient records associated with 161 disorders,[81] Rzhetsky and colleagues found that many complex diseases co-occur in individuals at rates significantly above or below those expected by random chance. The implication is that disorders that co-occur more often than expected share some causes, while disorders that co-occur less often than expected imply that one disorder protects against the another.

For disorders that are known to be genetic, higher than expected co-occurrence indicates shared genes that predispose to both diseases. Likewise genetic disorders that appear to preclude each other are likely to share genes which are causal for one disorder and protective for the other. Thus, for both over and under co-occurring disorders, genes associated with one disorder can be considered candidate genes for the other disorder. In this way, the large numbers of sequence variants in personal genomes can be reduced to manageably small numbers of candidate gene hypotheses.

If appropriate genome-wide expression data is available for large numbers of healthy and disease-carrying individuals, then expression differences between the two groups can be used to identify groups of genes associated with the disease state. For example, Watkinson *et al.*[82] identify pairs of genes whose expression can jointly (but not singly) distinguish between cancerous and healthy prostate tissue. The resulting network of gene pairs forms a prostate-cancer candidate-gene set.

Characterizing Environmental and Life-History Effects

The preceding discussions focused on interpretation of DNA sequence variants. How do we integrate genetic predispositions with current and past health status data and arrive at a diagnosis? The answer depends in part on the nature of biomarkers used to monitor and

---

[81] A Rzhetsky *et al.*, Probing genetic overlap among complex human phenotypes, *Proceedings of the National Academy of Sciences of USA*, 2007, **104**(28): 11694–11699.

[82] J Watkinson *et al.*, Identification of gene interactions associated with disease from gene expression data using synergy networks, *BMC Systems Biology*, 2008, **2**: 10.

characterize health status. In the sections below, we first discuss biomarker selection and then review approaches to the interpretation of biomarker data.

Identification of multi-parameter biomarkers. As noted in Chapter 4, biomarkers can take diverse forms. Because biomarkers are selected on the basis of their predictive performance, many only indicate the onset of a disorder without pinpointing its cause. For example, many biomarkers measure the downstream effects of a disorder.

Since biomarkers rarely directly measure the mechanistic origin of a disorder, they are seldom perfect predictors. In some cases, the prediction of a disorder will turn out to be false (false positives). In other cases, a disorder will not be detected (false negatives). Three commonly used measures of biomarker performance are sensitivity, false alarm rate (FAR), and positive predictive value (PPV), defined as:

$$\text{Sensitivity} = \frac{\text{number of true positives}}{\text{number of true positives + number of false negatives}}$$

$$\text{False Alarm Rate (FAR)} = \frac{\text{number of false positives}}{\text{number of true negatives + number of false positives}}$$

$$\text{Positive Predictive Value (PPV)} = \frac{\text{number of true positives}}{\text{number of true positives + number of false negatives}}$$

"True positive" and "true negative" refer to the number of positive and negative test results that are correct. Sensitivity measures the fraction of cases detected. FAR is the ratio of false-positive test results to the total number of healthy individuals tested. Finally, PPV measures the proportion of patients with positive test results who are correctly diagnosed.

Usually, a test can be performed at various levels of stringency. Improved sensitivity results in a higher FAR, while reducing the FAR reduces sensitivity. Plots of Sensitivity versus FAR characterize how well a test performs at all stringency levels. For historical reasons, they are known as Receiver Operating Characteristic or ROC curves. An example ROC curve is shown in figure 8.5. The dashed 45° line indicates the characteristic performance arising from random decisions. All useful tests will have ROC curves above this line.[83]

The area under a ROC curve is an indication of the effectiveness of the test. For random test results, the area is the triangle under the 45° line, i.e. 0.5. For a perfect test, we require

---

[83] For a review, see MH Zweig and G Campbell, Receiver-Operating Characteristic (ROC) plots: a fundamental evaluation tool in clinical medicine, *Clinical Chemistry*, 1993, **39**(4): 561–575.

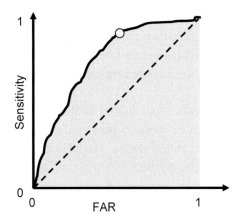

Figure 8.5: An example ROC curve.

that when sensitivity = 1, FAR = 0. Therefore the area under the curve will be equal to one. For real-life biomarker-based tests, the area under the ROC curve will be somewhere between these two bounds.

For medical tests, we require sensitivity to be very close to one, and FAR close to zero (i.e. an area under the curve close to one). The ROC characteristic curve illustrated in figure 8.5 would not be an adequate medical test, because for sufficiently high values of sensitivity, the false alarm rate will be unacceptably high (see for example the operating point marked by the white disk).

For screening scenarios, where large numbers of individuals are tested but few are expected to be affected, the False Alarm Rate is not a very informative statistic (because of the large number of true negatives). In such cases, the utility of a test is best characterized in terms of sensitivity and PPV instead.[84]

The take-away message from the above discussion is that all biomarker-based tests will have non-zero rates of false positives and misses. One way to improve the performance of tests is to use multiple test criteria. For example, weight alone is not a good marker for obesity. Using weight and height together, we are more likely to estimate a person's obesity correctly. Using additional markers such as waist to hip ratio, muscle mass and bone density may improve our prediction quality further.

---

[84] J Davids and M Goadrich, The relationship between precision-recall and ROC curves, *Proceedings of the 23rd International Conference on Machine Learning* (ACM International Conference Proceeding Series; Vol. 148), 2006, Pittsburgh, Pennsylvania, USA, pp. 233–240.

The usefulness of each additional test criterion used will depend on the extent to which it performs a function complementary to already-selected criteria. Feature selection and classifier design are important topics in computer science.[85] Here, it suffices to note that careful selection of multiple biomarkers can greatly improve the predictive value and robustness of tests.

The advantage of using multiple biomarkers is illustrated schematically in figure 8.6. Suppose the tested population can be divided into two classes: those affected ("×" symbols in the plot), and those not affected (filled disks). Hypothetical test 1 and test 2 scores for the two populations are indicated by the positions of the "×"s and disks in the scatter plot. Note how there is considerable overlap in individual test scores between the two populations. Thus, individually neither test can correctly classify all individuals. However, a weighted sum of the two test scores — represented by the dashed line — can perfectly distinguish between the affected and unaffected individuals.[86]

Given the considerable degree of inter-individual variability (see Chapters 3 and 4), it is crucially important that the performance of biomarker panels be cross-validated on data from populations not used during the design and development phase. In particular, the performance of biomarkers developed using data from a single genetic group or a single country may be significantly different for individuals from other genetic or environmental backgrounds.

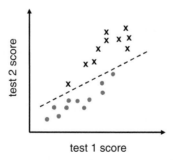

Figure 8.6:   Multiparameter classification.

---

[85] See http://www.support-vector-machines.org/ and http://www.kernel-machines.org/ for information on two popular approaches. Reviewed in A Ben-Hur *et al.*, Support vector machines and kernels for computational biology, *PLoS Computational Biology*, 2008, **4**(10): e1000173.

[86] The line can be described by the equation: **test 2 score = m • test 1 score + C**, where m and C are constants and • indicates multiplication. An individual is affected if **test 2 score > m • test 1 score + C**. Put another way, an individual is affected if $\frac{1}{C}$ • **test 2 score** − $\frac{m}{C}$ • **test 1 score** > 1.

pathway diagrams are used to summarize the known interactions of variants with other genes and drugs. The pathway diagrams are interactive, so that clicking on any component takes the user to textual information on the relevant drug, metabolite or gene. An example pathway diagram showing the role of the *ABCB1* gene in the Taxane pathway is shown in figure 9.3. Fifteen other drug-associated pathways are also listed in PharmGKB for *ABCB1*.

Not every physician will want to be confronted with this level of detail for every variant within a patient's genome. The important point to note here is that such information *can* be made readily available to physicians in the form of web-based interactive presentations where the physician can choose the level of detail presented through a series of drill-down menus.

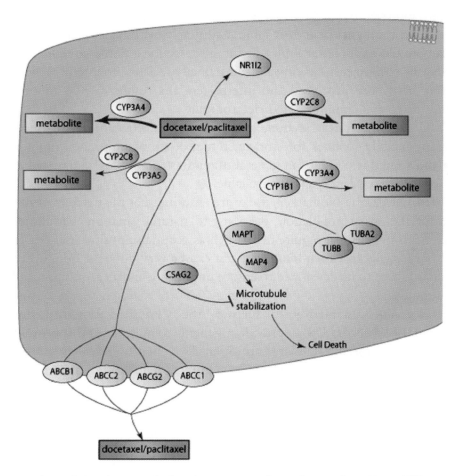

Figure 9.3: An example drug-interaction pathway diagram from PharmGKB.

Just as Promethease searches the SNPedia and compiles a custom report based on the results from commercial SNP arrays, future personalized genomics tools will be able to search databases such as PharmGKB and construct detailed reports of the disease-associated and drug-specific variants in individual patient's genomes. And just as PharmGKB already provides multiple types of resources (including interactive diagrams and links to other resources), future personal genomics reports will comprise multiple types of data organized in hierarchical and interactive forms tailored to meet the needs of physicians and other health professionals.

Interpreting Biomarker Data

Most clinically approved biomarker tests generate easy-to-interpret results. These results can be communicated to physicians in text tables and graphs within Electronic Health Records (see the example blood cholesterol plot in the next section). Some complex tests, such as imaging, typically require custom software and trained specialists to interpret the results. But once the data have been processed and interpreted, the results can be annotated by the specialist and stored and communicated electronically.

As we will see in the next section, many existing Electronic Health Record (EHR) systems already include facilities for storage, retrieval, annotation, and communication of image data and related text. Thus, in general, the infrastructure for widespread and routine clinical use of biomarker data is already available.

Major challenges facing biomarkers today are ensuring that the marker (or marker set) is appropriate for the target population, is robust to natural variations within and between people, and meets operational requirements such as cost, false-positive/false-negative rates, and patient burden. Because each biomarker test can be very different from others, typically custom standards have to be developed to ensure uniformity in data acquisition and the interpretation of results.

Existing regulatory standards — such as the Clinical Laboratory Improvement Amendments (CLIA)[8] in the US — provide a general compliance and approval framework for biomarkers. More specific standards are often developed by research and development communities in order to improve performance and meet regulatory requirements.[9]

---

[8] http://www.fda.gov/cdrh/clia/

[9] See for example AR Padhani *et al.*, Diffusion-Weighted Magnetic Resonance Imaging as a cancer biomarker: consensus and recommendations, *Neoplasia*, 2009, **11**(2): 102–125; MK Tuck *et al.*, Standard Operating Procedures for serum and plasma collection: Early Detection Research Network Consensus Statement Standard Operating Procedure Integration Working Group, *Journal of Proteome Research*, 2009, **8**(1): 113–117. See also http://edrn.nci.nih.gov/resources/standard-operating-procedures.

As we noted in Chapter 7, emerging molecular biomarker tests increasingly utilize measurements of multiple genes, mRNAs, proteins or metabolites. Such multi-variable biomarker panels typically rely on complex classification algorithms to translate a set of measurements into a clinical diagnostic. Efforts are currently underway to develop regulatory requirements for such algorithms.[10]

Electronic Health Records (EHRs)

The increasing sophistication of medical practice over the past century has led to a mushrooming of medical disciplines and specialties. Today, diagnosis and treatment for most serious conditions typically requires contributions from multiple specialists and large numbers of tests. Personal genomics and the emerging biomarkers for personalized medicine add further complexity to this already difficult task.

We noted the economic and personal costs of avoidable adverse reactions in Chapter 1. In addition to these, errors and inefficiencies in communications among individuals and departments (e.g. loss or mishandling of paper records) have been widely noted.[11]

Computerized systems that facilitate record keeping, communications, and data analysis offer an attractive solution to these challenges. By putting all data relating to each patient in a single easily accessible location, EHRs also make it easier for physicians to assemble a complete picture of the patient.

Some example EHR application areas and tasks are listed in Table 9.1. The top three application areas directly aid the physician, the patient, and their interactions, while the bottom four exploit the computerization of health records to deliver public health, research, and administrative benefits. Not all of these features need be implemented within any single software application or organization,[12] but each application can potentially improve both the quality and cost-effectiveness of medical care.[13]

---

[10] See for example the US FDA's Draft Guidance for *In Vitro* Diagnostic Multivariate Index Assays: http://www.fda.gov/cdrh/oivd/guidance/1610.pdf.

[11] See for example MAB Makeham, M Mira and MR Kidd, Lessons from the TAPS study: communication failures between hospitals and general practices, *Australian Family Practice*, 2008, **37**(9): 735–736; EG Poon *et al.*, "I wish I had seen this test result earlier!" — Dissatisfaction with test result management systems in primary care, *Archives of Internal Medicine*, 2004, **164**(20): 2223–2228.

[12] Indeed, it has been argued that a "mix and match" approach would be far more efficient. KD Mandl and IS Kohane, No small change for the health information economy, *New England Journal of Medicine*, 2009, **360**(13): 1278–1281.

[13] PM Kilbridge and DC Classen, The informatics opportunities at the intersection of patient safety and clinical informatics, *Journal of the American Medical Informatics Association*, 2008, **15**(4): 397–407.

Table 9.1: Example EHR application areas and tasks.

Application area	Example tasks
Patient records	Record entry tools patient controlled access privileges health history summaries
Treatment management	Order entry (prescriptions, tests) access to test results drug interactions and safety alerts
Decision support	Biomarker interpretation genome interpretation treatment selection
Public health	Notifiable disease reporting epidemic surveillance drug adverse effects surveillance
Research	Data anonymization and access tools programming interface for data mining
Data sharing and communication	Patient-physician medical team, specialist test centers insurance/provider
Administrative support	Billing appointments, referrals quality assurance indicators

Accompanying the advent of EHRs, there is an opportunity — and a need — to provide patient access to medical records, and to give patients the chance to ensure their records are correct, up-to-date, and available to those who need them. To date, three distinct types of Personal Health Records have been developed:[14]

(1) Patient portals into provider-specific EHRs. Typically such portals allow patient access to most components of a provider's EHR, but exclude specific parts such as progress notes. They may also provide tools for patients to update their records (e.g. immunizations, allergies), links to physician-recommended websites with information relating to their conditions, and secure email communications.

(2) Supra-institutional Personal Health Records also provide patient portals into EHRs. But instead of being clinic or provider specific, the records are patient-specific and shared by all care providers. This approach may be easier to adopt in countries with national healthcare systems. For example, in the UK the government has set out to develop a nationwide EHR provision for the entire National Health System (NHS). The £12 billion

[14] For a review of some US examples and challenges remaining, see JD Halamka, KD Mandl and PC Tang, Early experiences with personal health records, *Journal of the American Medical Informatics Association*, 2008, **15**(1): 1–7.

NHS National Programme for IT (NPfIT)[15] was started in 2002 and began pilot runs of its EHR system[16] in 2007. Although delays and financial over-runs have dogged the government-managed UK effort, it offers one important advantage over other systems: nationwide interoperability and access. As we will see below and in the next section, interoperability opens up opportunities for improved bedside-to-bench research, administration, and outbreak surveillance.

(3) Third-party hosted, Personally Controlled Health Records empower patients to integrate and maintain their full health records throughout life. This type of health record may prove particularly beneficial in the USA, where patients may need to switch healthcare providers when they change jobs.

In recent years, Microsoft and Google have both announced software tools for personally controlled health records,[17] and a group of eight large US companies[18] are sponsoring the development of a patient-controlled EHR system based on open-source software originally developed at the Boston Children's Hospital.[19]

Personally controlled health records put the burden of responsibility for maintaining complete and up-to-date records on the patient. On the other hand, institutional and supra-institutional EHRs limit the extent to which patients can manage their records.

Capital cost and organizational challenges have so far impeded the uniform and across-the-board adoption of EHRs.[20] Nonetheless, almost all industrialized countries now have programs aimed at universal provision of EHRs. Moreover, a number of highly successful EHR systems are already used widely and demonstrate the feasibility of EHRs.

---

[15] See the UK House of Commons, Committee of Public Accounts, Department of Health: The National Programme for IT in the NHS, Twentieth Report of Session 2006–07, HC 390, published April 2007. Available from www.parliament.the-stationery-office.com/pa/cm200607/cmselect/cmpubacc/390/390.pdf.

[16] http://www.connectingforhealth.nhs.uk/area

[17] See http://www.google.com/intl/en-US/health/tour/ and http://www.healthvault.com.

[18] Dossia is supported by AT&T, Applied Materials, BP USA, Cardinal Health, Intel, Pitney Bowes, Sanofi-Aventis and Wal-Mart. See http://www.dossia.org.

[19] IndivoHealth, from which Dossia is being developed, is licensed under the GNU Lesser General Public License (http://www.gnu.org/licenses/lgpl.html), which requires propagation of the open source license in all IndivoHealth based products. See http://www.indivohealth.org.

[20] AK Jha *et al.*, Use of electronic health records in U.S. hospitals, *New England Journal of Medicine*, 2009, **360**(16): 1628–1638; RJ Baron, Quality improvement with an electronic health record: achievable, but not automatic, *Annals of Internal Medicine*, 2007, **147**: 549–552.

To provide a flavor of EHR resources that will become universal in the near future, we will briefly review two exemplar hospital-based systems (VistA and WebOMR) below. Many additional commercial and non-commercial EHR software projects are underway. For example, the philanthropically sponsored OpenMRS system (http://openmrs.org/) is aimed at providing a free, open-source platform for EHRs in developing countries.

Notable among the commercial offerings is Microsoft's Amalga software suite, which in 2009 has three components:

(1) The business/clinical analytics module includes an aggregator and a set of analytic resources. The aggregator can be configured to automatically integrate data from all hospital departments (ranging from laboratory results, pharmacy orders, and clinical records to human resources, inventory, and billing). The analytic resources allow operations surveillance, data mining, and optimization of business processes.

(2) The hospital information management system allows all stakeholders (e.g. physicians and hospital managers) to access, analyze, and act upon appropriate portions of the integrated data.

(3) The life sciences research module provides a range of data analysis tools for personal genomics and personalized medicine research (the intention is to provide comprehensive data interpretation resources along the lines discussed in the previous chapter).

Besides its technical capabilities, the Microsoft Amalga system is notable in that it marks the entry of a software giant into healthcare IT, and highlights the increasing commercial opportunities in medical information management and analysis.

In contrast to the corporate-strategy roots of Amalga, VistA (the veterans health information systems and technology architecture) evolved within the US Department of Veteran's Affairs (VA) over several decades. The roots of Vista go back to projects started in the 1970s.[21] An earlier version of VistA was adopted across the VA health system in 1985. The current version of VistA is a comprehensive, full-featured and "industrial-scale" Health Information System.[22] We review some of its key features below.

---

[21] For a history of VistA, see SH Brown *et al.*, VistA — U.S. Department of Veterans Affairs national-scale HIS, *International Journal of Medical Informatics*, 2003, **69**: 135–156.

[22] The EHR portion of VistA, called the Computerized Patient Record System (CPRS), is freely available for download from http://www1.va.gov/cprsdemo. The full VistA software suite can be obtained by making a US Freedom of Information request: see http://www.hardhats.org/foia.html.

The VA is mandated to provide healthcare for about a quarter of the US population.[23] Its healthcare expenditure in 2009 is estimated at approximately $40 billion. The VA healthcare system currently includes 153 medical centers, 909 ambulatory care and community-based outpatient clinics, 135 nursing homes, 47 residential rehabilitation treatment programs, 232 Veterans Centers and 108 comprehensive home-care programs. Today, all of VA's healthcare providers use VistA. The figures 9.4 and 9.5 on the next two pages provide example views of VistA's Computerized Patient Record System (CPRS) user interface.

Figure 9.4 shows a snapshot of the patient summary sheet (called the CPRS cover sheet). The various sub-windows list the patient's personal details, "active problems", medications, allergies/adverse reactions, clinical reminders, lab results, "vitals", and appointments. The headings within each sub-window can be used to drill down to more detailed information. For example, the inset window (highlighted) shows details relating to the patient's Systolic Heart Failure record, accessed from the item list in the "active problems" sub-window.

At the bottom left-hand corner of the CPRS window is a series of ten tabs, providing links to sheets for summary data (the cover sheet being viewed), problems, medications, orders, notes, consultations, surgery, discharge summary, labs, and reports. Figure 9.5 shows a composite of some of the data that can be accessed through VistA. The top panel shows a portion of the "active problems" sheet (the inset shows the detailed data relating to the highlighted entry).

The panel at the bottom right shows an example of lab results presented as a graph. In this case the data represent cholesterol measurements over time. The two straight horizontal lines indicate minimum and maximum reference levels, allowing quick visual assessment of the data.

The bottom-left panel shows examples of the many types of medical images that can be stored, annotated and retrieved in VistA. Additional VistA features include billing and accounting tools, registration and eligibility tools, and data communication and sharing tools. Tools for surveillance and research are under development.[24] An open-source community-maintained version of VistA with additional features for use in new settings (called WorldVista[25]) was released in 2008.

---

[23] The number of people eligible includes veterans, family members or survivors of veterans. In 2008, about 5.5 million people received care through the VA. See http://www1.va.gov/opa/fact/vafacts.asp.

[24] For more details, see http://www.va.gov/vista_monograph.

[25] http://worldvista.org, see also http://www.hardhats.org

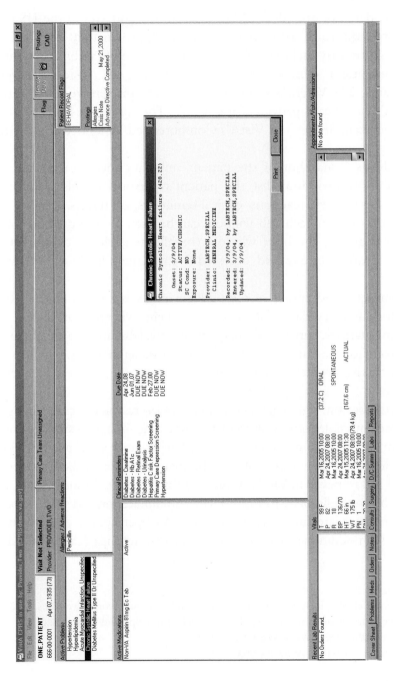

Figure 9.4: An example VistA patient summary sheet.

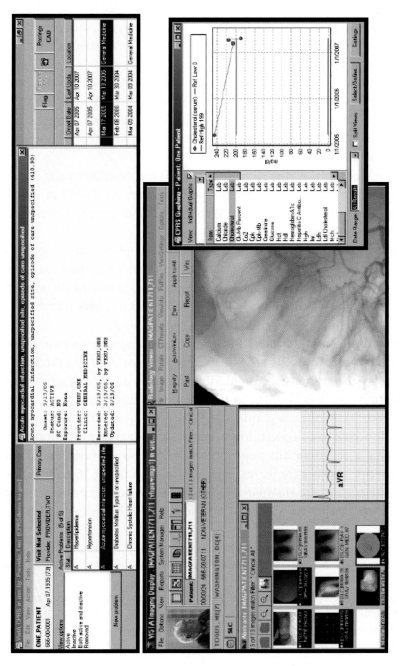

Figure 9.5: Example patient data in VistA.

In summary, VistA demonstrates that large-scale EHR systems spanning multiple geographically distributed institutions of different sizes and types, can successfully offer the kind of medical informatics infrastructure that will be necessary for effective translation of personal genomics and personalized medicine.[26]

The second exemplar EHR system that we will briefly review here is CareGroup's WebOMR[27] (web-based online medical records) system developed at the Beth Israel Deaconess Medical Center (BIDMC). In 2007, BIDMC had 585 licensed beds and over 5,000 staff, including 1,200 physicians.[28] The current, fully web-based version of WebOMR has been in operation at the BIDMC since 2003.

A key feature of WebOMR is that all patient records are centralized and patient-centered. Moreover, physicians can inspect not only the patient's current data, but also his/her past test results and medical history. And because WebOMR is web-based, it can be accessed from anywhere within the BIDMC's wireless network service.

There are two distinct interfaces to WebOMR: one is physician-centered, the other is patient-centered. The patient-centered interface has a series of tabbed sheets with features similar to the VistA CPRS. The physician interface provides access to the same patient data, but it is organized in terms of the physician's workflow and includes interfaces dealing with the physician's daily schedule, task list, and results notifications.

Like VistA, WebOMR has a number of integrated warning and decision support features. For example, if abnormal test results are not followed up within a specified time, an alert is automatically sent to the responsible physician. Potential adverse reactions to drugs and drug–drug interactions are also brought to the physician's attention.

In summary, WebOMR offers a good example of how web-based patient-centered EHR systems can provide access to complete, long-term patient data from any location with access to the internet. This is an important point because the growing numbers of sub-specialties and technology-specific services are increasingly making healthcare distributed and compartmentalized. Yet, the highly interdisciplinary nature of personal genomics and personalized medicine demand an integrated view of the patient. EHR systems such as VistA and

---

[26] For more on this topic, see P Longman, *Best Care Anywhere: Why VA Health Care is Better than Yours*, PoliPoint Press, 2007.

[27] WebOMR can be explored at: http://home.caregroup.org/webomr_training, CareGroup's home page is http://www.caregroup.org/.

[28] R Bohmer, FW McFarlan and J Adler-Milstein, Information Technology and clinical operations at Beth Israel Deaconess Medical Center, Harvard Business School case report number 9-607-150, 2007.

SimulConsult uses a Bayesian statistics approach to discover finding-diagnosis correlations directly from patient data. In principle, such pattern-matching methods could be used to mine patient data stored in large scale EHR systems, and expand CDS knowledgebases directly from patient records.

The above example CDS systems were aimed at diagnosis. Antibiotic selection aids go beyond diagnosis. On the one hand, they utilize patient-specific data to recommend additional diagnostic tests and targeted antibiotics.[45] On the other, they are combined with infection surveillance software[46] to identify the sources and types of cross-infections in hospitals. Such tools are particularly valuable in critical care units where physicians often have to make empirical decisions before detailed test results are available. For example, the TREAT antibiotic selection CDS system combines signs, symptoms, and laboratory tests with hospital surveillance data to predict the infectious agent (typically with comparable performance to experienced physicians).[47]

The third area in which task-specific CDS systems are already proving highly successful is patient-specific drug dosage selection. Personalized drug dosage selection is especially important for drugs with narrow therapeutic windows, and in children where there is greater variability among individuals. A good example is provided by the Pediatric Knowledgebase,[48] which provides custom web-based "dashboards" for different drugs. Each dashboard mines the hospital records for the drug of interest to construct a population dose-response model. The model is then used to predict patient-specific dosage requirements. The dashboard interface allows physicians to track model predictions against patient test results (see example image on the back cover of the book).

In addition to providing support for health professionals, computer-based decision support systems are also increasingly being developed to help patients understand and explore their condition. Many current patient decision aids simply provide explanatory information,[49] but more interactive and multi-faceted tools are becoming available. For example, the

---

[45] Reviewed in V Sintchenko, E Coierac and GL Gilbert, Decision support systems for antibiotic prescribing, *Current Opinion in Infectious Diseases*, 2008, **21**: 573–579.

[46] See for example http://www.theradoc.com/, http://tinyurl.com/SafetySurveillor, and http://www.carefusion.com.

[47] M Paul *et al.*, Prediction of specific pathogens in patients with sepsis: evaluation of TREAT, a computerized decision support system, *Journal of Antimicrobial Chemotherapy*, 2007, **59**(6): 1204–1207.

[48] http://stokes.chop.edu/programs/cpt/pkb/, JS Barrett *et al.*, Integration of modeling and simulation into hospital-based decision support systems guiding pediatric pharmacotherapy, *BMC Medical Informatics and Decision Making*, 2008, **8**: 6.

[49] For example, webMD: http://www.webmd.com.

diabetes PHD[50] service from the American Diabetes Association uses a patient-filled health status report to produce a personal diabetes susceptibility profile and predict risk levels for related events such as heart attack, stroke, kidney failure, and foot and eye complications. By changing the personal-choice variables in their health profile (e.g. stop smoking, lose weight, or take ACE inhibitors) users can see how making these changes will affect their long-term health.

The diabetes PHD patient decision support tool uses a highly detailed disease simulation model (Archimedes: see Chapter 8) to process detailed medical data (such as the number and types of medication the patient is on) to make its predictions. A variety of other molecularly detailed disease simulation models are being developed[51] and could be used in this way in future.

In the specific case of type 2 diabetes, it has been argued that eight simple clinical measures (age, sex, presence or absence of a family history, body-mass index, fasting glucose level, systolic blood pressure, high-density lipoprotein cholesterol level, and triglyceride level) can predict future incidence with 90% accuracy.[52] Thus a fully detailed simulation model may not be necessary in order for patients to explore the effects of lifestyle choices on their long-term health. For such scenarios, "system dynamics" modeling has been used in a number of settings to provide exploratory models.[53]

In systems dynamics, cause–effect relationships are typically first captured graphically in the form of a "box and arrow" influence diagram. A number of software aids are available to help translate influence diagrams into dynamic simulation models. The ease of model development using system dynamics modeling tools[54] can facilitate the translation of clinical expert knowledge into disease simulation models. We may therefore expect that many more ad hoc system dynamics disease models will be developed in the coming years as the need for greater patient participation in treatment planning grows.

---

[50] The "PHD" stands for personal health decision, available at: http://www.diabetes.org/diabetesphd.

[51] For examples, see http://mydigitalhealth.com, and K Thiel, Systems biology, incorporated? *Nature Biotechnology*, 2006, **24**(9): 1055–1057.

[52] KMV Narayan and MB Weber, Clinical risk factors, DNA variants, and the development of Type 2 Diabetes, *New England Journal of Medicine*, 2009, **360**(13): 1360.

[53] See for example JB Homer and GB Hirsch, System dynamics modeling for public health: background and opportunities, *American Journal of Public Health*, 2006, **96**(3): 452–458; GB Hirsch and JB Homer, Modeling the dynamics of health care services for improved chronic illness management, *Proceedings of the 22nd International Conference of the System Dynamics Society*, Oxford, England, July 2004.

[54] See for example http://www.vensim.com.

From Bench to Bedside and from Bedside to Bench

The preceding review of EHRs reveals two important trends in healthcare. First, the provision of EHRs and related IT resources will be comprehensive and nationwide in all industrialized countries in the near future. Second, EHRs will provide opportunities for large-scale data mining of health records (patient demographics, prescriptions, test orders and results, hospitalizations, reimbursement and billing, etc.).

We noted some of the benefits of large-scale EHR mining in the previous section. In this section, we take a more detailed look at the use of EHR mining to perform "observational studies" and pharmaco-epidemiology.

Observational studies offer a useful complement to clinical trials.[55] Clinical trials are typically expensive and short-term, involve relatively small numbers of people, and measure only surrogate end-points (rather than clinical outcomes). Observational studies on the other hand can involve millions of cases, use long-term medical records and measure actual clinical outcomes. They are supported by existing regulations such as the US Health Insurance Portability and Accountability Act of 1996 (HIPAA).[56] Moreover, observational studies based on analysis of EHRs can be very low-cost.

To date, two methodological issues have limited the usefulness of observational studies: confounding factors and selective reporting. Confounding factors arise from biases in the selection of case and control groups. For example, differences in insurance entitlements, or differences in services available in different states. Selective reporting is a form of confirmation bias, for example where a researcher repeatedly re-designs and re-runs analyses until results confirming expectations are produced.

Both of the above barriers will be overcome as large-scale, nationwide, and potentially multi-national EHR data become available for computational observational analyses. Confounding factors can be filtered out through randomized selection of cases and controls, as in clinical trials. Selection bias can be avoided by requiring that researchers pre-declare their aims and planned analyses at the time of requesting access to EHR data.[55]

The upshot of the ability to perform large-scale observational studies on EHR data is that translational medicine becomes bidirectional: from biological insight to clinical practice (as has been traditional), and from clinical data to new biological and medical insights (dubbed bedside-to-bench).

---

[55] J Avorn, In defense of pharmacoepidemiology — embracing the yin and yang of drug research, *New England Journal of Medicine*, 2007, **357**(22): 2219–2221.

[56] http://www.hhs.gov/ocr/privacy

In terms of classical pharmaco-epidemiology, the benefits of bedside-to-bench research have already been amply demonstrated in high profile cases such as delineating the adverse effects of Vioxx[37] and identification of suitable patient sub-populations for the cancer drug Iressa (gefitinib).[57] But observations at the bedside can also shed light on basic biology concepts and help clarify disease mechanisms.

The availability of a greater variety of biomarker tests, the increasingly quantitative nature of biomarkers, and the availability of personal genomes will accelerate bedside-to-bench discoveries. Nobel laureate Sydney Brenner may have intended to be provocative when he said "We don't have to look for model organisms anymore because we are the model organism."[57] But in the long run he is likely to be right, once again.

To go from the bedside to bench requires that clinical data be available for researchers to mine. This raises privacy issues. Traditionally, privacy has been maintained by de-identifying records. However, as more and more detailed data are recorded per patient, it becomes increasingly possible to re-identify patients from unique combinations of features in their records.

Re-identification would be considerably more difficult if the various components of each medical record were disassociated from each other. However, such an approach would also greatly reduce the usefulness of the records for bedside-to-bench studies. In the long term, it may be more realistic to limit *acting* on personal data, rather than trying to stop re-identification. In the near term, there is a clear and urgent need to debate and develop legal and organizational frameworks to address these issues (discussed in the next chapter).

Ceci n'est pas une pipe

This chapter has reviewed how clinical use of personal genomics and personalized medicine can be facilitated through information technology, and how the use of such technology will in turn spur new discoveries and greater personalization of medicine.

In *How Doctors Think* (Houhgton Mifflin, 2007), Jerome Groopman MD argued that increasingly narrow specialization, as well as time, financial and administrative pressures encourage physicians to reduce diagnosis to boiler-plate pattern-matching algorithms and decision trees. Using a wide variety of case studies from many branches of medicine, Groopman showed clearly that open-minded, case-by-case, ad hoc analysis of each patient's data and history can save lives, improve health outcomes, and provide greater job satisfaction for physicians.

---

[57] See H Ledford, The full cycle, *Nature*, 2008, **453**: 843–845.

We emphasize that sophisticated medical informatics tools are not intended to *replace* physicians' good practice habits. They simply facilitate the management and analysis of increasingly large-scale and detailed data. How successfully electronic health records, decision support systems, and other computational resources are exploited will depend on the extent to which they aid good practice habits.

CHAPTER 10

Organizational, Legal and Ethical Challenges

Medical practice has always had to deal with complex legal and ethical questions. The widespread adoption of personal genomics and personalized medicine in the coming years will pose a host of new ethical, legal, social and organizational challenges. This chapter reviews some of the key issues relating specifically to personal genome sequencing and personalized medicine.

Given that personal genomics and personalized medicine are undergoing enormous change at a very rapid pace, it may seem premature to discuss ethical and regulatory frameworks. However, given the large amounts of genomic and biomarker samples and data that are currently being generated and stored, there is a pressing and urgent need to put in place legal and ethical support systems that protect individuals while facilitating research and maximizing commercial opportunities.

Privacy Considerations

It is surprisingly easy to identify a person's genetic profile from de-identified sequence data. Between 30 and 80 independent DNA markers (e.g. SNPs) are sufficient to identify an individual uniquely and with high confidence.[1] If some of the markers are rare, fewer are needed, and if relevant information about a subject's medical history, ethnic background, or the genomic sequences of blood relatives is known, a few independent markers may be enough. Data obfuscation techniques do not improve the situation significantly.[1]

Computer security measures can reduce the probability of inappropriate access to data. However, given the many thefts and accidental releases of personal data to date (including tax data, social security numbers, credit card details, clinical trial data and electronic health records)[2]

---

[1] Z Lin, AB Owen and RB Altman, Genomic research and human subject privacy, *Science*, 2004, **305**: 183.
[2] See for example J Kaiser, NIH reports breach of patient records, *Science*, 2008, **319**: 1746. For a European example, see http://tinyurl.com/UKdataTheft.

it may be necessary to put in place additional legal barriers to unauthorized exploitation of personal genomic and medical records.

At present, enacted laws in the EU and US only partially address all medical privacy concerns.[3] In the EU, there is considerable variation among member countries in local laws and the implementation of the European Union Directive on Data Protection.[4] In the US, the key relevant laws are the Protection of Human Subjects (PHS) code of federal regulations (2005 revision),[5] the Health Insurance Portability and Accountability Act (HIPAA, 1996–2000),[6] and the Genetic Information Nondiscrimination Act (GINA)[7] of 2008.

PHS is concerned with the use of information collected for research purposes. It does not apply to data "[…] if the information is recorded by the investigator in such a manner that subjects cannot be identified, directly or through identifiers linked to the subjects."[5] Thus, anonymized research data (that could potentially allow re-identification due to unique feature sets) are not specifically regulated by PHS.

GINA specifically prevents health insurers and employers from discriminating on the basis of genetic information. HIPAA is the most general of the three laws. It includes a Privacy Rule about the types of use and disclosure of personally identifiable health information. However, a February 2009 report[8] by the Institute of Medicine (IOM) Committee on Health Research and the Privacy of Health Information found that:

> "…the HIPAA Privacy Rule does not protect privacy as well as it should, and that, as currently implemented, the Privacy Rule impedes important health research. The committee found that the [HIPAA] Privacy Rule
>
> (1) is not uniformly applicable to all health research,
> (2) overstates the ability of informed consent to protect privacy rather than incorporating comprehensive privacy protections,
> (3) conflicts with other federal regulations governing health research,
> (4) is interpreted differently across institutions, and
> (5) creates barriers to research and leads to biased research samples, which generate invalid conclusions."

---

[3] WW Lowrance and FS Collins, Identifiability in genomic research, *Science*, 2007, **317**: 600–602.

[4] http://ec.europa.eu/justice_home/fsj/privacy/law/

[5] http://www.hhs.gov/ohrp/humansubjects/guidance/45cfr46.htm

[6] http://www.hhs.gov/ocr/privacy/

[7] The full text of GINA is available from http://tinyurl.com/GINA-text.

[8] SJ Nass, LA Levit and LO Gostin (Editors), Beyond the HIPAA Privacy Rule: Enhancing Privacy, Improving Health Through Research, available at http://www.nap.edu/catalog.php?record_id=12458.

In addition, HIPAA only applies to health plans, healthcare clearinghouses and healthcare providers. Other organizations that handle patient data, for example those providing personally controlled health records such as Dossia, Microsoft, and Google (discussed in Chapter 9) are not subject to HIPAA.

Informed consent is often the key mechanism through which researchers obtain access to genomic and health records.[9] It can be argued that consent should be obtained for each new research use of previously obtained data.[10] Moreover, genetic data reveal information not only about the individual giving consent, but also about relatives and future offspring (see also issue (2) raised by the IOM report above).

Should family members have the right to veto the release of a person's genetic profile to third parties (e.g. researchers)? Should they be informed if significant genetic risk is discovered in the course of research?[11] Because of the heavy burden such considerations impose on researchers, alternative approaches that allow re-use of data while maintaining anonymity are needed.[12]

One solution to privacy management would be to route all data queries through a "trusted third-party" (TTP) organization. An example of such a service in the US is SureScripts[13] (formerly RxHub), which retrieves patient prescription histories for hospitals seeking medication reconciliation. The advantage of this approach is that the TTP can be regulated to high standards and maintain audit trails for all transactions. However, the need to go through a third party would complicate research and may have significant inhibitory effects on research.

Another approach is anonymization. The most straightforward approach to anonymization is to remove all personal identification details from records permanently. But such an

---

[9] For a fuller discussion of informed consent issues, see J Smith-Tyler, Informed consent, confidentiality, and subject rights in clinical trials, *Proceedings of the American Thoracic Society*, 2007, **4**: 189–193. Also JE Lunshof *et al.*, From genetic privacy to open consent, *Nature Reviews Genetics*, 2008, **9**: 406–411.

[10] AL McGuire *et al.*, Ethical, legal, and social considerations in conducting the Human Microbiome Project, *Genome Research*, 2008, **18**(12): 1861–1864; KM Boyd, Ethnicity and the ethics of data linkage, *BMC Public Health*, 2007, **7**: 318.

[11] For examples of such dilemmas, see G Chan-Smutko *et al.*, Professional challenges in cancer genetic testing: who is the patient? *Oncologist*, 2008, **13**: 232–238. See also the recommendations in AL McGuire, T Caulfield and MK Cho, Research ethics and the challenge of whole-genome sequencing, *Nature Reviews Genetics*, 2008, **9**: 152–156.

[12] For a discussion of the inadequacy of informed consent as a means of ensuring privacy, see P Taylor, When consent gets in the way, *Nature*, 2008, **456**: 32–33.

[13] http://www.surescripts.com/additional-services.html

approach would limit the reusability of the data. An alternative is to assign a unique identifier to each individual, but only allow highly regulated access to the table that matches individuals to their unique identifiers. While more burdensome, this approach has the advantage of allowing more flexible re-use of the data. For example, a recent study in Scotland used encrypted identifiers to merge health records and census data, thus providing more comprehensive but de-identified individual records.[14]

The challenge with all anonymization approaches is that they do not guard against possible re-identification. For instance, until recently, data from SNP-based genome-wide association studies were made public in aggregate (and therefore de-identified) form. But it was recently demonstrated that such aggregate data can reveal whether a subject was associated with a disease if the individual's SNP profile is known.[15] As a result, several funding agencies have now removed the aggregate data from public websites.[16]

Unlike credit card and other information, genetic data remain correct and valid for the entirety of a person's life and beyond. In particular, knowledge of parents' genomic data can be used to infer their descendant's genetic characteristics, raising the possibility of unforeseen and unpredictable misuses of genetic data in the distant future.

Copies of genomic data produced by consumer genetics firms and commercial sequencing facilities, as well as health records maintained on third-party servers (including patient-controlled health records), could potentially remain in the company's electronic archives indefinitely. Moreover, when an individual moves to another country or changes healthcare providers within the same country, their genomic and health data could potentially remain in archives at the place of origin, thus proliferating privacy risks. For these and other reasons, it has been argued that — by default — the life-expectancy of all electronically held personal data should be limited.[17]

Even if data erasure is legally mandated, some leakage of personal genomic and health information may be unavoidable. For example, a recent study in Canada found that about 10% of second-hand personal computers had significant amounts of personal health

---

[14] CM Fischbacher *et al.*, Record linked retrospective cohort study of 4.6 million people exploring ethnic variations in disease: myocardial infarction in South Asians, *BMC Public Health*, 2007, 7: 14.

[15] N Homer *et al.*, Resolving individuals contributing trace amounts of DNA to highly complex mixtures using high-density SNP genotyping microarrays, *PLoS Genetics*, 4(8): e1000167.

[16] J Couzin, Whole-genome data not anonymous, challenging assumptions, *Science*, 2008, 321: 1278.

[17] J-F Blanchette and DG Johnson, Data retention and the panoptic society: The social benefits of forgetfulness, *Information Society*, 2002, 18(1): 1–13.

information on their hard disk drives.[18] Thus additional laws may be needed to guard against the exploitation of accidentally released personal genomic and health data.

Accuracy and Usefulness of Tests

As discussed in Chapter 8, tests are rarely prefect predictors. Rather, they strike a balance between Sensitivity, False Alarm Rate (FAR), and Positive Predictive Value (PPV). Moreover, a test that strikes a good balance in one population or setting may be less than optimal in another.

Failure to detect a disorder is clearly undesirable. The financial and personal (e.g. emotional) costs associated with false positives also impose a requirement for very low false alarm rates. Tests with inadequate FAR and PPV typically have to be supplemented with additional tests, resulting in increased costs.

In addition to financial, time and resource-usage costs, some tests may incur health penalties (e.g. biopsies), or may provide little or no health benefit. Thus, there is a need to characterize the performance of tests rigorously. The US Centers for Disease Control have adopted four criteria for the evaluation of clinical tests.[19]

Analytic validity refers to the accuracy with which a marker (e.g. a DNA sequence or the abundance of a molecule) is measured. Analytical validity is a measure of technical/ technological capability.

Clinical validity refers to the ability of a test to detect or predict a clinical condition (for example a disease) based on the measured quantity. For example, a disease allele with low penetrance has low clinical validity as a stand-alone test.

As with analytic validity, the clinical performance of a test can be characterized as a balance between Sensitivity, FAR and PPV. Similarly, a test may be deemed inadequate in itself, but useful in conjunction with additional observations and tests.

Clinical utility addresses the extent to which test results will be clinically useful. For example, a test that accurately predicts a disease for which no interventions are available and no time of onset can be predicted would be of little clinical utility. Moreover, some tests with high

---

[18] K El Emam *et al.*, An evaluation of personal health information remnants in second-hand personal computer disk drives, *Journal of Medical Internet Research*, 2007, 9(3): e24.
[19] http://www.cdc.gov/genomics/gTesting/ACCE.htm

analytic and clinical validity may have risks that outweigh the potential health benefits of a diagnosis.

Impediments and safeguards refer to social, ethical, legal and organizational factors that affect the availability or usefulness of a test. For example, some high-cost tests or treatments may simply not be available through government-subsidized healthcare systems or some private insurers.[20] In the US, there is considerable geographical variation in government healthcare provision,[21] and poorer people are typically under-diagnosed and under-treated,[22] while others are often over-treated.[23]

Because of the complexity of results, psychological impact on the patient, or potential impact on society, it may be necessary to require that some test results are discussed with a counselor or other professionals. The additional cost and complexity may deter some patients.

Whole genome sequencing is not a test in itself. However, the *interpretation* of specific portions of a genome can have clinical validity and utility. At present, there are no overarching laws regulating the quality of genomic and biomarker tests, but specific laws regulate particular applications. For example, in the US, the Food and Drug Administration (FDA) regulates commercially sold test kits as "medical devices".[24] However, tests developed and offered by a single laboratory are not included in this category and are thus not regulated by the FDA. Furthermore, FDA approval of a test kit for a particular indication does not prohibit its use for other indications.

In some cases, the majority of the uses of a test may be "off label". For example, in the case of a CYP2D6 and CYP2C19 genotyping chip from Roche,[25] the FDA clearance simply states that the test "may be used as an aid to clinicians in determining therapeutic strategy

[20] For an example cost–benefit analysis, see N Simpson *et al.*, The cost-effectiveness of neonatal screening for Cystic Fibrosis: an analysis of alternative scenarios using a decision model, *BMC Cost Effectiveness and Resource Allocation*, 2005, **3**: 8.

[21] For Medicare spending, these issues are reviewed in the 2008 Congressional Budget Office report: Geographic variation in health care spending, available at http://www.cbo.gov/doc.cfm?index=8972.

[22] See for example DC Miller *et al.*, Prostate cancer severity among low income, uninsured men, *Journal of Urology*, 2009, **181**: 579–584.

[23] See DC Miller *et al.*, Incidence of initial local therapy among men with lower-risk prostate cancer in the United States, *Journal of the National Cancer Institute*, 2006, **98**(16): 1134–1141; S Brownlee, *Overtreated — why too much medicine is making us sicker and poorer*, Bloomsbury USA, 2007.

[24] http://www.fda.gov/cdrh/clia/

[25] http://www.amplichip.us/

and treatment dose for therapeutics that are metabolized"[26] by the two genes. All clinical uses of this test are therefore technically "off label" and not regulated.[27]

For the particular case of direct to consumer genetic testing, in the absence of appropriate laws, the American Society of Human Genetics published a statement[28] in 2007 stressing the need for quality assurance, clinical validity of tests, and user education.

Ethical Challenges

Most of the ethical challenges facing personal genomics and personalized medicine arise within established medical practice already. Such issues include the "medicalization" of some conditions previously considered natural human variation[29] (e.g. mild forms of depression or attention deficit hyperactivity disorder), direct-to-consumer sales and marketing of services (e.g. CT scans), and commercial interests biasing research findings[30] and regulatory reviews.[31] Since these issues are common to all forms of medicine and are studied widely, they will not be discussed in general terms here. However, some specific comments are necessary.

A key aspect of medicalization of conditions such as depression or obesity is that often people with very different conditions are lumped together under a single clinical label. For example, psychotic depression, melancholic depression, severe depression, and chronic depression are distinct from each other and also distinct from undifferentiated depressive symptoms.[32] Personalized medicine can help this situation by stratifying diagnoses and allowing specific treatments where intervention is necessary, beneficial and available.

---

[26] www.amplichip.us/documents/CYP450_P.I._US-IVD_Sept_15_2006.pdf

[27] SH Katsanis, G Javitt, and K Hudson, A case study of personalized medicine, *Science*, 2008, **320**: 53–54.

[28] K Hudson *et al.*, ASHG statement on direct-to-consumer genetic testing in the United States, *American Journal of Human Genetics*, 2007; **81**(3): 635–637.

[29] See for example D Healy, The latest mania: selling bipolar disorder, *PLoS Medicine*, 2006, **3**(4): e185.

[30] See for example JS Ross *et al.*, Guest authorship and ghostwriting in publications related to rofecoxib — a case study of industry documents from rofecoxib litigation, *Journal of the American Medical Association (JAMA)*, 2008, **299**(15): 1800–1812.

[31] D Carpenter *et al.*, Drug-review deadlines and safety problems, *New England Journal of Medicine*, 2008, **358**(13): 1354–1361.

[32] RT Mulder, An epidemic of depression or the medicalization of distress? *Perspectives in Biology and Medicine*, 2008, **51**(2): 238–250.

Direct-to-consumer sales of genetic tests have raised considerable concern in recent years with respect to issues such as accuracy of tests, clinical validity and utility, and the impact the results may have on the recipient.[33] In the case of nutritional genomics, there is an additional concern that unregulated tests may be used to sell clinically unproven "nutritional supplements".[34]

When used for minor or non-clinical conditions, consumer genetic and other tests can be seen as extensions of longstanding and widespread patterns of self-medication. For example, in the UK, non-prescription (over the counter) drug sales in 1994 exceeded one third of the National Health System's prescribed drugs bill.[35] Excluding cases arising from lack of access to healthcare (e.g. due to financial and geographical barriers), most self-medication appears to be for minor or well-characterized ailments and so may not pose a significant health hazard.[36]

Consumer genetic and other tests can also motivate the people who take them to modify risky behaviors, to identify symptoms early on, and to seek professional help. For example, dietary changes can delay the onset of cardiovascular diseases[37] and type 2 diabetes[38] in patients with specific genotypes.

In general, it should be noted that personal genomics and personalized medicine do not imply or require direct-to-consumer services. Indeed, we have assumed throughout this book that all tests will be ordered and interpreted by professionals within appropriate clinical settings.

Another common ethical challenge with specific implications for personalized medicine concerns "moral hazards" associated with "expert services". Moral hazards arise when incentives to behave in a less than ideal manner are not counter-balanced by disincentives. For example, cheap or free medical coverage may encourage greater uptake of more expensive

---

[33] See for example G Naik, As gene tests spread, questions follow, *Wall Street Journal*, December 13, 2007, p. D1; JL Fox, What price personal genome exploration? *Nature Biotechnology*, 2008, **26**(10): 1105–1108.

[34] NM Ries1 and D Castle, Nutrigenomics and ethics interface: direct-to-consumer services and commercial aspects, *OMICS*, 2008, **12**(4): 245–250.

[35] A Blenkinsopp and C Bradley, Patients, society, and the increase in self medication, *British Medical Journal (BMJ)*, 1996, **312**: 629–632.

[36] See for example L Grigoryan *et al.*, Self-medication with antimicrobial drugs in Europe, *Emerging Infectious Diseases*, 2006, **12**(3): 452–459.

[37] JM Ordovas, Genetic interactions with diet influence the risk of cardiovascular disease, *American Journal of Clinical Nutrition*, 2006, **83**(suppl): 443S–446S.

[38] LR Ferguson, Dissecting the nutrigenomics, diabetes, and gastrointestinal disease interface: from risk assessment to health intervention, *OMICS*, 2008, **12**(4): 237–244.

(but not necessarily better) brands of drugs.[39] Similarly, cheap or free genomic analyses and biomarker tests may encourage demand for unnecessary tests.

The expert services problem refers to transactions where the customer is poorly equipped to judge the quality of advice or service being offered, as is typical for patients. Although patients may seek a second or third opinion on major issues, they do not generally have the technical expertise to compare the clinical validity of tests, the accuracy of the interpretation of multiple test results, and whether they are being under or over-treated.

When there is no good way for the "customer" to assess the quality of advice or service they receive, providers face a moral hazard. They may benefit from acting in ways that are not in their patients' best interest. For example, they may profit from recommending more expensive or more convenient tests/treatments, or from over-treating patients (to satisfy the demands of anxious patients, to give the impression of providing a high level of care, or to guard against complaints/litigation).[40]

Personal genomics and personalized medicine could also increase the vulnerability of patients and physicians to marketing[41] because the complexity of the issues will make it more difficult to make a well-informed, rational choice among alternatives. This applies not only to overt advertising, but also indirect factors such as biases in the publication of research findings and clinical trial results.[42]

Yet another general issue which becomes more acute with personalized medicine is the ownership of tissue samples and test data. We touched on this issue in terms of privacy, but there are additional ethical and legal implications. For instance, can a blood sample submitted for a particular test also be used by the testing laboratory for additional, unrelated analyses (e.g. for research)?[43] Should the police be allowed to search personal genome and biobank databases for matches to evidence found at crime scenes?[44]

---

[39] See for example N Pavcnik, Do pharmaceutical prices respond to potential patient out-of-pocket expenses? *RAND Journal of Economics*, 2002, **33**(3): 469–487.

[40] See for example KK Fung, Dying for money: overcoming moral hazard in terminal illnesses through compensated physician-assisted death, *American Journal of Economics and Sociology*, 1993, **52**(3): 275–290.

[41] Chapter 9 (pp. 203–233) of *How Doctors Think* (Jerome Groopman, Houghton Mifflin, 2007) provides an excellent summary of these issues.

[42] See for example EH Turner *et al.*, Selective publication of antidepressant trials and its influence on apparent efficacy, *New England Journal of Medicine*, 2008, **358**(3): 252–260.

[43] JA Robertson, The $1000 genome: ethical and legal issues in whole genome sequencing of individuals, *American Journal of Bioethics*, 2003, **3**(3): W35–W42.

[44] T Caulfield *et al.*, Research ethics recommendations for whole-genome research: consensus statement, *PLoS Biology*, 2008, **6**(3): e73.

These considerations highlight the need for new laws and ethical guidelines. They also suggest a need to provide both patients and physicians with confidential and independent professional advice. We will return to this concept in the next section.

The Need for Patient Participation in Personalized Medicine and its Challenges

Much of medical practice today already requires that patients participate in making complex decisions. Here is an example described in *Intern: A Doctor's Initiation* (by Sandeep Jauhar, Farrar, Straus and Giroux, 2008, p. 232):

> "I told him he had five blockages in three arteries and two options. Angioplasty could open the arteries without surgery, but he would need two procedures, one of which could be started right away. Open heart surgery, on the other hand, probably offered him the best chance of not having to undergo another procedure in the future. "But it's a big surgery," I added."

Needless to say, the patient did not know what to make of his choices.[45] To date, a wide variety of question-prompt sheets, counseling and coaching methods, and web-based (sometimes interactive) resources have been developed to help patients make sense of their medical data, arrive at a clear understanding of their options, and make informed decisions based on personal preferences.[46]

Personalized medicine will increase the complexity of decision making by virtue of the added information it provides. At the same time, because it offers predictive genetic markers, earlier diagnosis, and better staging of diseases, personalized medicine will offer patients more prevention, amelioration, and containment alternatives than current practices.

We noted in the previous section that both physicians and patients can sometimes be influenced by marketing claims, biased data availability and other manipulations. In the case of patients, there is the added problem that they may have little or no scientific background.

As with the previous section, these considerations suggest a need for independent and unbiased sources of data and advice for both patients and physicians. A large part of such

---

[45] A more complex example is given in *ibid*. pp. 220–224.

[46] See for example D Stacey, Rv Samant and C Bennett, Decision making in oncology: a review of patient decision aids to support patient participation, *CA Cancer Journal for Clinicians*, 2008, **58**: 293–304. See http://decisionaid. ohri.ca/ for a list of over 200 decision-aid web-based information resources. See http://tinyurl.com/AD-decision for an example decision-aid addressing the question: Should I take medications to treat Alzheimer's disease?

services could be provided through internet-based, pre-packaged media such as web pages, databases, and interactive exploratory and educational tools. As we noted earlier, the technologies for such resources are already well established. The challenge here will be to provide timely, authoritative and trustworthy summaries of the latest findings in formats suitable for people from diverse backgrounds.

At the other end of the spectrum, for major medical decisions and complex case-histories, there will be a need for a new generation of independent (fiduciary) medical counselors who can guide patients through the interpretation of complex test data and explore the long-term benefits and drawbacks of alternative interventions. Physicians too may need such fiduciary advice in order to navigate the complex daily legal and ethical challenges arising from individual cases. We will discuss the education of such counselors in the next chapter.

In between the above two extremes of support for personalized medicine is the emerging trend towards telemedicine, and internet-based consultations carried through email and "live" video teleconferencing. Such services have the advantages of extreme cost-effectiveness and flexibility. Among other things, arranging appointments can be automated, queries can be automatically routed to the most appropriate expert, and waiting times minimized. Telephone and internet-based consultations have attractive features for both providers and patients, and are therefore likely to become increasingly popular for non-critical and routine consultations.

Economic Considerations

Healthcare is becoming increasingly costly. Consider the cost of drugs as an example. As illustrated in figure 10.1,[47] over the past quarter century the cost of developing new drugs has increased approximately ten-fold, while the number of new products (based on new and existing molecules) has not changed significantly.

In part because of the increased development costs, new drugs have become increasingly expensive to purchase. Combined with the effects of social factors such as greater wealth, more health consciousness, greater availability of healthcare services, and an aging population, the amount spent per person per year on pharmaceuticals has been increasing at near

---

[47] This figure is reproduced from FL Douglas and L Mitchell, in US Department of Health and Human Services report: Personalized Health Care: pioneers, partnerships, progress, November 2008, p. 112. Available from http://www.hhs.gov/myhealthcare/news/presonalized-healthcare-2008.html.

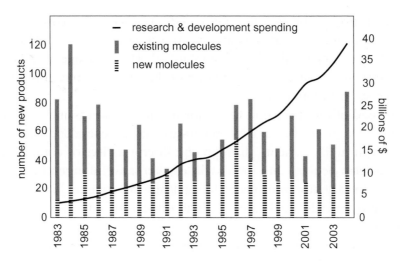

Figure 10.1: The increasing cost of pharmaceutical research and development.

exponential rates since the 1960s. Figure 10.2 illustrates this trend for a few representative developed countries (note the logarithmic scale).[48]

Drugs are not the only part of healthcare that is becoming more costly. Cost increases are also due to more sophisticated diagnostic technologies (e.g. CAT/PET/MRI scans), advanced surgical procedures (e.g. robotic radio-surgery[49]), and better implantable devices (e.g. neurostimulation devices[50] for pain management and epileptic seizures; cardiac rhythm management systems). These more sophisticated diagnostic and intervention technologies also create a need for highly trained (and therefore more expensive) specialists.

Overall, increases in total healthcare expenditure are outpacing increases in wealth. Figure 10.3 shows healthcare expenditure as a percentage of the Gross Domestic Product (GDP) for a few representative developed countries.

Though there are clear differences between nations, the long-term underlying trend is similar in all cases. In the US, health expenditure as a percentage of the GDP is expected to

---

[48] This figure and the next (10.3) are based on data from the Organisation for Economic Cooperation and Development (OECD) 2008 data: http://tinyurl.com/OECD-health-data (accessed February 2009).
[49] See for example http://www.cyberknife.com.
[50] See for example http://tinyurl.com/medtronic-neurostim.

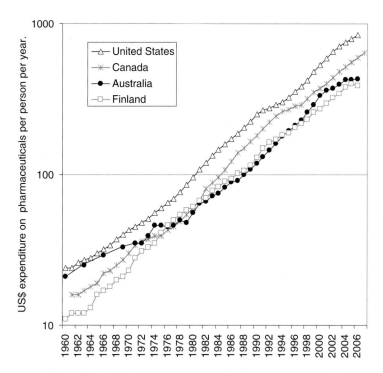

Figure 10.2: Increasing expenditure on pharmaceuticals per person per year.

grow at an average rate of 1.9% per year in the foreseeable future, reaching about one-fifth of the GDP in 2017.[51]

How will the emergence of personal genomics and personalized medicine be affected by and affect these trends? Will soaring healthcare costs lead insurers and governments to resist personalized medicine? Or will personalized approaches lead to cost savings over whole lifetimes, and thus offer value for money? Will people in the developing countries be increasingly left behind? And will poorer people in industrialized countries be able to benefit from personal genomics and personalized medicine?

At present, there is not enough practice-based evidence to answer these questions fully. However, several lines of indirect evidence suggest that the answers to these questions can be very positive.

---

[51] US DHHS national health expenditure projections 2007–2017. Available from http://www.cms.hhs.gov/ NationalHealthExpendData/Downloads/proj2007.pdf.

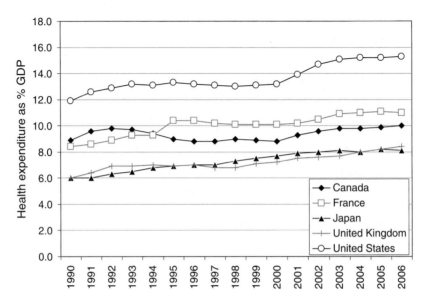

Figure 10.3: Increasing national expenditure on healthcare.

With respect to developing nations, there is considerable diversity in infrastructure and healthcare needs. Some of the emerging economies (e.g. Brazil, India) have strong industrial and research communities. At the other end of the spectrum, annual per capita healthcare expenditure in sub-Saharan countries is only about US $30.[52]

The coordinated international response to the current malaria, tuberculosis, and HIV/AIDS epidemics offers insights into how advances in biotechnology can address the needs of the less developed nations. Some key observations are:

- When appropriately funded, public–private partnerships such as The Global Fund to Fight AIDS, Tuberculosis and Malaria[53] can dramatically improve health outcomes in developing nations.
- Grants to researchers in developed countries aimed at the development of appropriate solutions for developing nations (e.g. through the Global Fund,[53] and the Bill and

[52] B-J Hardy *et al.*, The next steps for genomic medicine: challenges and opportunities for the developing world, *Nature Reviews Genetics Supplement: Genomic medicine in developing countries*, 2008, **9**(1): s23–s27.
[53] http://www.theglobalfund.org/en/

Melinda Gates Foundation[54]) are an effective means of exploiting advanced technologies and the latest research insights to deliver targeted solutions.

- At present rates of occurrence, prevention is far more critical (and cost-effective) than treatment. This is true for both infectious and also non-infectious diseases (e.g. those arising from factors such as exposure to toxic materials and poor nutrition). However, for at least another decade, prevention alone will not be sufficient. Early detection will be essential both for improved health outcomes and also for containment of costs arising from ill health.

- Appropriate technologies are needed that enable (a) low-cost, point-of-care screening with high analytical and clinical validity; (b) same-day delivery of test results, counseling services, and treatment plans; and (c) stratified diagnosis (e.g. identification of multi-drug-resistant and extensively drug-resistant TB strains, identification of latent TB, especially TB in people with HIV).

New cheap, easy-to-run genetic and microfluidic assays (see examples in Chapter 7) tailored to the needs of under-developed and remote regions are already emerging. For example, with the support of the Gates Foundation, a simple, low-cost, and rapid test for cervical cancer has been demonstrated in India[55] and China.[56] The test is currently being developed and "productized" by the multinational company Qiagen. To quote from the Qiagen press release:[57]

> "The *care*HPV test can be conducted by workers with minimal healthcare training and education. Once collected, samples of vaginal or cervical cells are prepared for analysis using a kit of reagents that contains its own water supply. The testing itself is conducted on easily portable equipment and will run on batteries."

The emergence of "appropriate technology" devices such as *care*HPV suggests that funding availability and peer-recognition can effectively drive the development of advanced technologies suitable for less-developed regions.

---

[54] http://www.gatesfoundation.org

[55] R Sankaranarayanan *et al.*, HPV screening for cervical cancer in rural India, *New England Journal of Medicine*, 2009, **360**: 1385–1394.

[56] Y-L Qiao *et al.*, A new HPV-DNA test for cervical-cancer screening in developing regions: a cross-sectional study of clinical accuracy in rural China, *Lancet Oncology*, 2008, **9**(10): 929–936.

[57] Available at http://tinyurl.com/careHPV, accessed March 2009.

New technologies ranging from microfluidics to cell phones are enabling some developing nations to adopt advanced healthcare practices at a fraction of the cost of established medical infrastructure in the West. For example, a recent report by the United Nations Foundation lists 51 telemedicine projects in developing countries around the world.[58] In India, the Apollo group has been developing telemedicine strategies and multimedia electronic health record systems since 1999,[59] and in Mexico, MedicallHome offers phone-in consultations[60] as a low-cost health service.

African, Latin American and East Asian populations comprise genetically diverse groups.[61] While current initiatives aimed at people in developing regions treat entire nations as homogenous populations, the availability of cheap and easy tests will ultimately enable the use of genetically stratified preventive strategies and focused screening of at-risk populations. To this end, national population genotyping projects are already underway in several developing countries, including Mexico,[62] India,[63] and Thailand.[64]

For economically disadvantaged groups within industrialized countries, success will also depend critically on policy decisions at national and international levels. Because of their highly targeted nature, stratified — and ultimately personalized — medicines are sometimes compared to "orphan drugs" for rare diseases.[65]

The development of orphan drugs is strongly supported by economic incentives in the EU, US and elsewhere.[66] In the US, the Orphan Drugs Act (ODA) was enacted in 1983, and by May 2008 had resulted in the introduction to the market of 325 approved drugs.[67]

[58] Available at http://tinyurl.com/mHealth-report.

[59] http://www.apollohospdelhi.com/apollo-group/group-companies.html

[60] See http://www.medicallhome.com.mx.

[61] M Jakobsson *et al.*, Genotype, haplotype and copy-number variation in worldwide human populations, *Nature*, 2008, **451**: 999–1003.

[62] G Jimenez-Sanchez *et al.*, Genomic medicine in Mexico: Initial steps and the road ahead, *Genome Research*, 2008, **18**: 1191–1198. See also http://www.inmegen.gob.mx.

[63] Indian genome variation consortium, Genetic landscape of the people of India: a canvas for disease gene exploration, *Journal of Genetics* (of the Indian Academy of Sciences), 2008, **87**(1): 3–20.

[64] See http://www4a.biotec.or.th/thaisnp/.

[65] Orphan drugs are those aimed at diseases affecting less than 200,000 people in the US, or less than 5 in 10,000 people in the European Union.

[66] See http://www.fda.gov/orphan/oda.htm and http://www.emea.europa.eu/htms/human/orphans/intro.htm.

[67] FDA Cumulative list of all Orphan Designated Products that have received Marketing Approval: http://tinyurl.com/FDA-orphanDrugs.

severity of the disease, the personal and financial costs of intervention, the patient's other health needs, the effectiveness of lifestyle changes, etc.[83] Thus, the appropriate time and mechanism of testing and intervention has to be tailored to each individual. The patient will need to make a personal decision (informed by physicians and counselors) as to when a threshold is crossed, and what action should be taken.

Consider this example. As discussed in Chapter 1, Ashkenazi Jewish women with a family history of ovarian and breast cancer and a mutation in their *BRCA1/2* genes have a more than 85% chance of developing breast cancer at some point. If you are a young Ashkenazi woman whose mother and aunt both had breast cancer, should you be tested for *BRCA* mutations? If the result indicates you have the mutation, should you have a mastectomy? Immediately? Or after having children? These are complex ethical and personal issues. Their resolution is well beyond the scope of this book, but they have been explored in detail elsewhere.[84]

Prolonging Life versus Promoting Quality of Life

As figure 10.5 illustrates, there has been an extraordinary improvement in human life expectancy over the past century.[85] Note how the life expectancy versus income curves for different years saturate at different heights. The saturation points indicate where basic needs such as nutrition, sanitation, and primary healthcare (e.g. vaccinations, antibiotics) are met.

For example, a person earning the 1991 equivalent of $4,800 per year[86] in the USA in 1900 had an average life expectancy of just 49 years. In 1990, a person with the same equivalent income had an average life expectancy of 71 years (ibid.), an approximately 45% improvement.

Note how, beyond the saturation point, the 1990 curve does not become completely flat. Rather, greater wealth provides longer lives.

In recent years, there has been an ongoing debate that the latest medical procedures and drugs have transformed acute diseases into chronic ones. This observation has led to questions about the extent to which medical advances should be used to prolong life irrespective of the quality of the extended life.[87]

---

[83] For a review, see JP Evans, C Skrzynia and W Burke, The complexities of predictive genetic testing, the *British Medical Journal (BMJ)*, 2001, **322**: 1052–1056.

[84] See for example M Gesssen, *Blood Matters: From Inherited Illness to Designer Babies, How the World and I Found Ourselves in the Future of the Gene*, Houghton Mifflin Harcourt, 2008.

[85] Source: World Bank, World Development Report 1993: Investing in Health, Washington DC, 1993.

[86] Earnings outside of USA are expressed in terms of 1991 "local purchasing power".

[87] See for example R Martensen, *A Life Worth Living: A Doctor's Reflections on Illness in a High-Tech Era*, Macmillan, 2008.

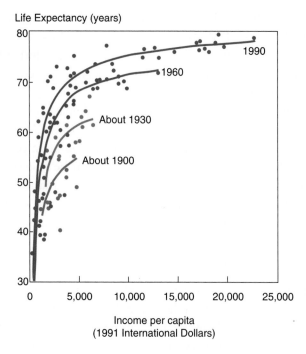

Figure 10.5: Dependence of life expectancy on income over the past century.

Judging "quality of life" is difficult and subjective. As with the other issues discussed in this chapter, the better diagnostics and the wider range of potential intervention strategies provided by personal genomics and personalized medicine will intensify the ongoing debate. However, it should be noted that, rather than prolonging suffering, early detection through personal genomics and personalized medicine can improve the affected person's quality of life by allowing the adoption of disease prevention strategies such as:

- Lifestyle changes, e.g. targeted exercises, dietary changes, stress reduction, avoidance of toxic exposures, etc.
- Preventive drugs (e.g. aspirin for people at risk of stroke) and vaccines (e.g. the angiotensin II vaccine for high blood pressure[88]).

---

[88] The angiotensin II vaccine is currently in clinical trials; see AC Tissot *et al.*, Effect of immunisation against angiotensin II with CYT006-AngQb on ambulatory blood pressure: double-blind, randomised, placebo-controlled phase IIa study, *Lancet*, 2008, **371**: 821–827.

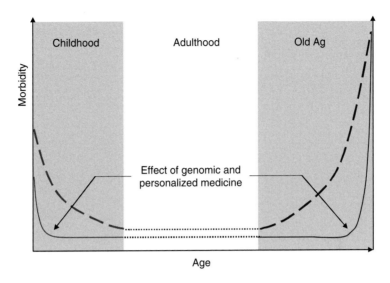

Figure 10.6: Hypothesized morbidity trends before and after the introduction of personalized medicine.

- Surgery (e.g. angioplasty) and medical devices that reduce the likelihood of future severe adverse events (e.g. defibrillators).

As summarized in the schematic figure 10.6, morbidity tends to be highest in early childhood and old age (dashed curves, arbitrary scales). Personal genomics and personalized medicine can reduce morbidity rates in both phases (solid curves). For example, as noted in Chapter 1, children born with the metabolic disorder PKU can avoid brain damage by adopting a strict diet following early diagnosis. In old age, for diseases that cannot be cured, it is likely we will able to delay the onset of the disease or severe symptoms (as with current HIV treatments).

CHAPTER 11

Coda — Engineering the Future of Medicine

As we saw in the introduction to this book, the confluence of technological innovations, societal demands, and better understanding of how genetic and environmental factors affect health is bringing about a paradigm shift in medicine.

We noted in Chapter 2 that cellular function is determined by the interactions of thousands of gene products, and saw in Chapter 3 that there is considerable variation in the genetic make-up of individuals, even among people with a shared genetic background. Such differences make each of us biochemically unique, and endow each individual with differing strengths and susceptibilities. Lifestyle and environmental factors operate on this genetic landscape somewhat like a topiarist shaping trees and shrubs (discussed in Chapters 4 and 5).

The technologies to enumerate our genomic individuality (reviewed in Chapter 6), to quantify the effects of environmental influences on cellular and organ function (explored in Chapter 7), and to predict health outcomes (the subject of Chapter 8) are beginning to reach the market now and will become widely available in the coming decade. The computational infrastructure to help physicians and patients explore large amounts of personal data is already being adopted around the world (reviewed in Chapter 9), and discussions of legal and practical frameworks for personalized medicine are under way (see Chapter 10).

In many ways, genomically personalized medicine is here already. In February 2009, the US FDA created a new position in its Office of Chief Scientist: Senior Genomics Advisor. The announcement states: "We stand ... on the brink of a new era of personalized medicine."[1]

Over the past decade, drugs and other interventions tailored to the needs of specific populations are becoming increasingly common. Table 11.1, reproduced from a September 2008 report[2]

---

[1] http://tinyurl.com/FDA-Genomics-Advisor
[2] Priorities for Personalized Medicine, Report of the President's Council of Advisors on Science and Technology, September 2008. Available from http://www..ostp.gov/galleries/PCAST/pcast_report_v2.pdf.

Table 11.1: Ten examples of personalized medicine products.

Product	Company	Technology/Test type	Application/Disease
HER2/neu tests	Several	Two types of test are available: immunohistochemical tests measuring expression of the HER2/neu protein and FISH tests measuring amplification of the HER2/neu gene	Determine eligibility of breast cancer patients for treatment with Herceptin® (trastuzumab)
Trofile™ assay	Monogram Biosciences	Uses cultured cell lines to assess the interaction of the patient's HIV-1 strain with different cell-surface receptors (phenotype)	Determine eligibility of HIV patients for treatment with Selzentry™ (maraviroc)
TPMT assays	Several	Two types of test are available: measuring the presence of TPMT gene variants or the level of TPMT enzyme activity	Set dose of thiopurine drugs to maximize therapeutic efficacy while minimizing bone marrow toxicity in diseases such as acute lymphocytic leukemia, inflammatory bowel disease, and severe active rheumatoid arthritis
Invader® UGT1A1 assay	Third Wave Technologies	Uses PCR to measure presence of UGT1A1*28 gene variant (genotype)	Set dose of irinotecan in colorectal cancer patients to maximize therapeutic efficacy while minimizing side effects of diarrhea and reduced white blood cell count
AlloMap® test	XDx	Uses quantitative PCR to measure expression of 20 genes, algorithm to convert results to quantitative composite score (multivariate genotype array)	Identify heart transplant patients at low risk for acute cellular rejection, may allow reduced use of biopsy for monitoring and/or more precise tailoring of immunosuppressive regimen

(Continued)

Table 11.1: (*Continued*)

Product	Company	Technology/Test type	Application/Disease
Oncotype DX®	Genomic Health	Uses quantitative PCR to measure expression of 21 genes, algorithm to convert results to quantitative composite score (multivariate genotype array)	Quantifies the risk of systemic recurrence and assesses the value of chemotherapy in patients with newly diagnosed, early stage invasive breast cancer
Antiretroviral drug resistance tests	Many	Wide variety of both genotypic and phenotypic tests commercially available	Assess presence of drug-resistant HIV strains to enable selection of effective antiretroviral regimen
AmpliChip® CYP450 test	Roche Diagnostics	Uses PCR amplification and DNA microarray technologies to assess presence CYP2D6 and CYP2C19 gene variants	Inform dosing decisions for a range of drugs that are metabolized to differing extents by variants of the CYP2D6 or CYP2C19 isoenzymes
Warfarin metabolism tests	Many	Variety of kit and laboratory implementations to assess presence of CYP2C9 and VKORC1 gene variants	Inform warfarin dosing decisions in patients requiring anticoagulation therapy
HLA B*5701 test	Many LDTs	Generally use PCR amplification and sequence-specific oligonucleotide probes to assess presence of B*5701 allele	Identify HIV patients likely to suffer severe hyper-sensitivity reaction to the antiretroviral drug abacavir

by the President's Council of Advisors on Science and Technology, lists ten examples of personalized medicine applications already available.

In this concluding chapter, we discuss some of the implications of these rapid developments and highlight some key opportunities and challenges that lie ahead.

Engaging and Educating the Stakeholders

Possibly the greatest challenge facing personal genomics and personalized medicine is how physicians and patients/customers can be educated and empowered to make well-informed choices. New educational programs are needed and must be started immediately if they are to educate a significant proportion of health professionals and the general public within a decade.

Support groups, online discussion forums, wikis, and blogs will continue to be important ways in which physician clubs and patient groups can exchange information, experiences, and recommendations. Assuming that the social, legal and technical challenges will be addressed, four additional measures can help:

1. **Data-sharing among providers, and between providers and researchers** will enable rapid analysis of large-scale data and its translation into medical benefits (e.g. identification of sub-populations for whom a drug is particularly effective or ineffective). It will also facilitate the practice of evidence-based medicine, and provide patients and physicians with up-to-date data on the effectiveness of different interventions.

   Such data sharing can only happen if appropriate data and (electronic) communication standards are established early on (see Chapter 9), and if privacy and other patient rights (Chapter 10) are addressed thoroughly.

   It will be difficult to reconcile propriety formats and practices once they become established. It is therefore imperative that action be taken now to bring stakeholders together and address these issues. Governments and funding agencies can create the right conditions for the emergence of data sharing standards by raising awareness of its benefits across all sectors of healthcare.

2. **Independent reviews** of the literature, new treatments, etc. are already widely used (for example, Cochrane Reviews[3]). Currently, the difficulty of assessing the quality of analysis/advice provided limits the usefulness of information provided by non-governmental

---

[3] http://www.cochrane.org/reviews/

organizations. One solution pursued by some EHR providers (Chapter 9) is to give physicians the ability to recommend third-party websites within the patient's personal health records. As technologies for assessing author trustworthiness and contribution relevance improve, patients and physicians will be better able to exploit third-party resources.

3. **Continuing education of physicians** in the use of latest technologies is increasingly essential as monitoring and intervention technologies become ever more complex. Physicians trained and certified in specific sub-specialties, and physicians who routinely perform complex procedures often outperform non-specialists. For example, a recent study found that implantable cardioverter-defibrillator (ICD) procedures performed by electrophysiologists were less likely to lead to complications than when the same procedure was performed by cardiologists.[4]

   Continuing physician education is a subject fraught with concerns about costs, and the biasing effects of industry sponsorship. Collaborative arrangements between professional bodies, universities, and funding organizations will be necessary to develop independently funded and certified continuing education programs for physicians and other health professionals.

4. **The creation and education of "fee-only" medical counselors**. Fee-only fiduciary financial advisors are widely used to avoid and bypass the expert services problem and other moral hazards in personal finance. A similar need is increasingly emerging as medicine becomes more personalized.

   Genetic counseling is currently a requirement for certain tests and optional for others. As large numbers of early-detection health tests become available, the need for genetic and non-genetic counseling will increase. Patients are going to need help understanding not only the biology of their condition, but also the test technologies and intervention available, and the nature and significance of statistical predictions.[5]

   Issues of who pays the counseling bills and how providers judge the necessity and usefulness of counseling services will likely pose some challenges. Nonetheless, the creation and education of fee-only personalized medicine counseling services may be unavoidable in order to provide patients/clients with independent advice in complex cases.

---

[4] JP Curtis *et al.*, Association of physician certification and outcomes among patients receiving an implantable cardioverter-defibrillator, *Journal of the American Medical Association (JAMA)*, 2009, **301**(16): 1661–1670.

[5] As Stephen J Gould famously argued in a 1985 *Discover* magazine article entitled "The median isn't the message", statistical measures such as means, medians and risk ratios can mislead rather than help patients arrive at decisions.

There is currently a shortage of clinical geneticists and genetic counselors in most countries. For example, in Canada in 2007 (population ~30 million), there were only 218 medical geneticists, and only about 80 of these were practicing clinically. Newly qualified genetic counselors were graduating from Canadian universities at a rate of just 18 per year.[6] Similar patterns are observed in the USA and elsewhere.[7] For example, in the US, there are currently only ~30 genetic counseling training programs nationwide, producing a total of about 250 graduates each year.[8]

To address the need for more genetic and personalized medicine counselors, it will be necessary to increase investment in educational programs, and provide incentives for geneticists to remain in clinical practice.[9]

Moving Beyond the "One Treatment Fits All" Approach

The traditional approach to biomedical research has focused on animal models of human diseases. Once a drug, device, or other intervention has been shown to be effective in animals, it is tested on humans in a series of increasingly large, costly, and time-consuming clinical trials. In terms of drug development, the cost, complexity, low success rate, and long turnaround times associated with this approach have led pharmaceutical companies to focus on blockbuster drugs with sales potential exceeding $1 billion per year.

As we noted in Chapter 1, blockbusters have two fundamental limitations. Firstly, no two patients have exactly the same genetic susceptibilities or environmental exposure histories, so it is unlikely that their disorders will be identical. As a result, a small but finite proportion of

---

[6] A Silversides, The wide gap between genetic research and clinical needs, *Canadian Medical Association Journal (CMAJ)*, 2007, **176**(3): 315–316.

[7] For example, in the USA in 2003 (population about 300 million), there were only 610 clinical geneticists who spent at least 5% of their time on direct patient care. See JA Cooksey *et al.*, The medical genetics workforce: an analysis of clinical geneticist subgroups, *Genetics in Medicine*, 2006, **8**(10): 603–614; and JA Cooksey *et al.*, The state of the medical geneticist workforce: findings of the 2003 survey of American Board of Medical Genetics certified geneticists, *Genetics in Medicine*, 2005, **7**(6): 439–443. A 2007 count, and a map showing the geographical distribution of medical geneticists in the US is presented at http://thepersonalgenome.com/2007/12/shortage-of-geneticists-in-the-united-states.

[8] G Lorge, You decoded, *Stanford Alumni Magazine*, May/June 2009 issue, available at http://www.stanfordalumni.org/news/magazine/2009/mayjun/features/genetics.html.

[9] CJ Epstein, Medical genetics in the genomic medicine of the 21st century, American *Journal of Human Genetics*, 2006, **79**: 434–438.

patients receiving a blockbuster drug will either not benefit from it or have unacceptable levels of adverse reactions.

The second limitation of blockbusters is they tend to dominate the share prices and profits of pharmaceutical companies. When a blockbuster in-the-making fails a clinical trial, or when a blockbuster already in the marketplace is discovered to have unacceptable adverse effects, the maker's stock typically falls in double digit percentages. We noted a case involving GlaxoSmithKline in Chapter 1 — figure 11.1 is another example:[10] an approximately 40% fall in Merck's share prices when the company withdrew its arthritis drug Vioxx in October 2004 (see region highlighted in red).

One effect of such dramatic fluctuations in share prices is that it discourages long-term investment, strategic thinking and planning. Well after a product has reached the marketplace, the parent company may have to suddenly change plans, make cuts in research and development budgets, and lay off workers. Not surprisingly, then, pharmaceutical companies can sometimes be slow to accept bad news about a candidate drug, potentially putting patients at risk.

The above arguments are in favor of moving *away from* blockbuster drugs. But such a move will only be feasible if there are attractive alternative business models. Targeted, pathway-specific drugs — allowing individualized combination therapies — offer one such alternative.

The increased awareness of inter-individual differences has so far primarily motivated the development of population-specific drugs such as those listed in table 11.1 at the beginning

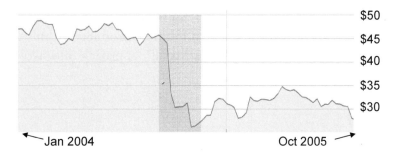

Figure 11.1:   Merck share prices before and after the company withdrew Vioxx.

---

[10] The figure shows New York Stock Exchange prices for Merck reproduced from Google Finance: http://tinyurl.com/Vioxx-MerckShares.

of this chapter. However, as we noted in Chapter 3, current evidence suggests that there is as much variation within populations as between them.

As diagnostic technologies become cheaper, more precise, and less intrusive, our appreciation of inter-individual differences will increase, creating a drive towards more personalized medicine.

Irrespective of genomic and life-history variations, there is one level at which we are all identical: we all share the same cellular pathways and processes. While the same pathology may impact people of diverse genetic and environmental backgrounds differently, all pathologies ultimately result in a set of dysregulated cellular pathways.

Instead of grouping patients in terms of gross symptoms, improved diagnostic technologies allow us to characterize each patient's specific condition in terms of particular combinations of dysregulated pathways. Tailored combinations of pathway-specific drugs can then provide highly personalized prescriptions. Viewed in this light, drugs and other medical interventions may be likened to clothing: we all wear *some* clothing, but we mix and match clothing items to suit our different body types and lifestyles/environments.

Another motivating force towards pathway-specific interventions is that the advent of personal genomics and medicine will create a virtuous circle: from bench-to-bedside and from bedside-to-bench. The more we use personal genomes and molecular biomarkers to characterize disorders in individuals, the more we will be able to stratify diseases — and their treatments — in terms of disordered pathways. Custom cocktails of pathway-specific drugs can then address each patient's unique combination of dysregulated pathways.

Because they will be used in different combinations for a large variety of conditions, pathway-specific drugs will have large markets. At the same time, because multiple pathway-specific drugs are likely to be needed for most complex diseases, manufacturer earnings will be more stably distributed across multiple products. In this way, drug development can progress beyond the "one-drug-fits-all" approach and avoid the pitfalls of blockbusters.

Are We There Yet?

Medicine has come a long way since the days of Hippocrates, Galen and Avicenna. Yet virtually all segments of society currently face enormous medical challenges. Taking the United States as an example, three trends are worthy of special mention here. Firstly, as we noted in Chapter 1, on average the world population is growing older.

In industrialized countries, the number of people aged 65 and over is increasing at a faster rate than the total population. For example, between 1950 and 2005, the total

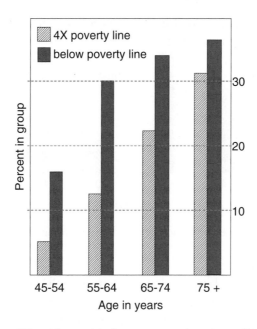

Figure 11.2: US residents with three or more chronic conditions in 2005.

population of the United States increased at an average annual growth rate of 1.2%. During the same period, the number of people 65 years and older grew ~2% per year. The number of people 75 years and over grew faster still, at an average rate of 2.8% per year.[11]

Figure 11.2[11] shows the number of US residents aged 45 or older suffering from three or more chronic conditions (2005 data). As expected, the proportion of sufferers increases with age, and the proportion of poor people with multiple chronic conditions is higher.

A surprisingly large proportion of people over 55 years old suffer from multiple ailments, even those who are financially secure (>10% above 55, >20% above 65, >30% above 75). Thus, the US population is getting older, and a significant proportion of older Americans have three or more chronic conditions. The total disease burden of the nation is therefore growing. Similar observations can be made for other industrialized countries.

The second noteworthy trend is that the above figures do not include many additional undiagnosed conditions. For example, figure 11.3 shows the percentage of US residents in

---

[11] Data from the US National Center for Health Statistics, Health, United States 2007, with Chartbook on Trends in the Health of Americans, 2007, Library of Congress Catalog Number 76–641496.

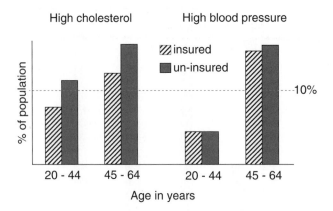

Figure 11.3:   Undiagnosed high cholesterol and hypertension in US patients during 2005.

2005 who were found to have high cholesterol (on the left) or hypertension (on the right), but who were not aware of the fact.[11]

For people over 45 years of age, the proportion of undiagnosed cases is fairly large: about one in eight people. Note that the proportions are only slightly less for people with health insurance, i.e. the lack of awareness is only partially due to lack of access to medical services.

As we develop more sensitive tests and understand mechanisms of diseases better, it is likely we will discover many more examples of undiagnosed conditions in the general population. With more widespread use of biomarkers, conditions that predispose to disease can be identified and addressed earlier. Through analysis of past patient data, we will also be able to better characterize factors that predispose to disorders as well as factors that can be used as early indicators of disease onset.

The third trend of special note here is that some conditions are becoming more prevalent in the general population, even among younger people. Figure 11.4 shows the proportion of US residents overweight or obese over the last four decades.[11] Note the alarming and remarkably similar rates at which the percentage of overweight children and the percentage of obese adults are both increasing. Since obesity is associated with increased risks for many disorders, these trends are indicative of greater rates of future health issues.

Medicine is necessarily a cautious and conservative discipline. Numerous obstacles can block or delay the translation of biomedical research to human studies, and the translation of

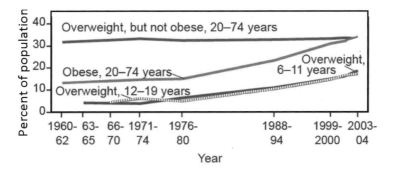

Figure 11.4: Proportions of overweight and obese US residents.

new clinical knowledge into everyday practice.[12] Indeed, more than four fifths of promising biological findings never reach clinical practice.[13] Biomedical breakthroughs that successfully make the transition to clinical practice often take more than 15 years to go from initial success to widespread adoption.[14] In the interim, adoption tends to occur in geographic, demographic, and sub-specialty islands.[14]

Although the pace at which technological innovations are introduced into the marketplace is increasing exponentially,[15] the complexity of healthcare suggests that universal use of personal genomics and personalized medicine is unlikely to occur within a decade. Does this mean that current work in personal genomics and personalized medicine is confined to basic research or niche markets? The answer is emphatically no.

The type and quality of healthcare that will be available in 15 years' time depends crucially on how medical research progresses *today*. Moreover, like most technological advances, adoption of personal genomics and personalized medicine will likely proceed in waves; enthusiasts and early adopters today, boutique health providers tomorrow, large-scale provision next, and finally universal access.[15] Apart from the long-term pay-off for companies that enter

---

[12] Reviewed in NS Sung *et al.*, Central challenges facing the national clinical research enterprise, *Journal of the American Medical Association*, 2003, **289**(10): 1278–1287.

[13] DG Contopoulos-Ioannidis, EE Ntzani and JPA Ioannidis, Translation of highly promising basic science research into clinical applications, *American Journal of Medicine*, 2003, **114**(6): 477–484.

[14] JM Westfall, J Mold, and L Fagnan, Practice-based research — "Blue Highways" on the NIH Roadmap, *Journal of the American Medical Association*, 2007, **297**(4): 403–406.

[15] See for example chapters 1 and 2 of Ray Kurzweil's *The Singularity Is Near*, Penguin Books, 2005.

the field early on, the size of the early-adopter and boutique provider markets is likely to be substantial and therefore more than self-supporting.

An important benefit of a gradual adoption of personal genomics and medicine is that it will allow time for the necessary social, legal and organizational infrastructure to evolve as the general public becomes familiar with the technologies and concepts.

There are good reasons to believe that personal genomics and personalized medicine will be adopted sooner and more broadly than previous medical advances. One reason is that personalized medicine may actually cut healthcare costs by providing better diagnostics, more effective treatments, and fewer episodes of severe illness.

Another factor that may lead to rapid uptake of personal genomics and personalized medicine is that the potential financial gains from mining quantitative, longitudinal patient data for bedside-to-bench research may far outweigh any extra costs associated with personal genomics and medicine. For example, healthcare cost savings could be made by identifying individuals with particular nutritional or lifestyle needs, or people who are particularly susceptible to specific environmental factors.[16] Such practices are already common for people at risk for conditions such as diabetes, heart disease and osteoporosis. Personal genomics and personalized medicine will simply expand the range of costly conditions that we may be able to guard against.

The combination of bedside-to-bench data mining and personal genomics/personalized medicine may also have a fortuitous benefit for the pharmaceuticals and biotechnology industries. Many drugs that have proven to be ineffective or have exhibited unacceptable adverse effects in mixed-population clinical trials may prove to be both effective and safe for specific sub-populations with a particular genomic background or health profile. Since there are many more drugs that fail clinical trials than succeed, companies already have large numbers of candidate molecular entities. Many of these have passed toxicity and other tests and can be repositioned at relatively low cost.

Finally, as noted in earlier chapters, personal genomes need only be sequenced once but can be mined for an individual's entire life. Thus, health insurance providers may come to view personal genomes as more cost-effective than multiple targeted genetic tests ordered over the lifetime of an individual. Indeed, the potential financial benefits of personal genome sequencing and personalized medicine have led to the suggestion that in the near future personal genomes may be sequenced for free as an integral part of

---

[16] For an example of the emerging field of toxicogenomcis, see AP Davis *et al.*, The Comparative Toxicogenomics Database facilitates identification and understanding of chemical-gene-disease associations: arsenic as a case study, *BMC Medical Genomics*, 2008, **1**: 48.

healthcare plans[17] (in much the same way that cell phone handsets are often provided free of charge when people sign up for long-term service plans).

Once a threshold is crossed and individual genomes are widely sequenced, the widespread availability of personal genomes would in turn drive the development of genome analysis tools and services (in the same way that application programs are written for some cell phones), which will make personal genomics more effective, and speed up the adoption of personalized medicine.

Let the Revolution Begin

During the 1980s and early 1990s, it became increasingly clear that there was an exponentially widening gap between manufacturing capacity and design productivity in the microelectronics industry, as illustrated in the schematic figure 11.5.[18]

Available microelectronic technology was increasingly under-exploited. The semiconductor industry responded by diversifying. Computing became personal, and then ubiquitous

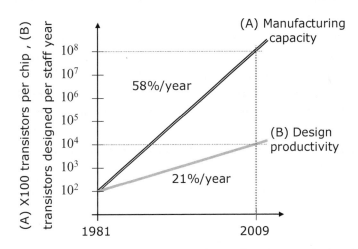

Figure 11.5: Divergence of design productivity and manufacturing capacity in microelectronics.

---

[17] Attributed to George Church (Harvard Medical School) in the *Economist*, 18th–24th April 2009, p. 11. Available online at http://tinyurl.com/free-genomes.

[18] Figure adapted from the SEMATECH 1997 International Technology Roadmap for Semiconductors report: http://www.sematech.org/.

and networked. At the same time, small, cheap microchips led to a blooming of consumer electronics (laptops, cell phones, digital cameras, programmable appliances, hand-held GPS, etc.) and the "dot com" boom of the late 1990s followed.

The life sciences are currently undergoing a similar transition. A burst of new technologies is providing increasingly sensitive, comprehensive, and quantitative views of the molecular systems underlying cellular and organ function and dysfunction. Witness the emergence of low-cost "omics" technologies, sensitive small-sample microfluidic assays, *in vivo* imaging, regenerative medicine, intelligent long-term implantable devices, key-hole imaging-guided robotic surgery, nano-medicine, and gene therapy.

How will these new and emerging technological capabilities be exploited in the health-care market? There is a ground shift towards highly individualized diagnosis and intervention, and the emergence of new health-related industries (e.g. consumer genetics and genomics, consumer biomonitoring devices and systems, telemedicine, and elective surgery).

In a manner reminiscent of the revolution in microelectronics, a radically different kind of medical practice is emerging. Just as computing has moved away from centralized main-frame computing and towards distributed (pervasive) low-cost and personalized resources, medicine is becoming personalized, and as it does so it is increasingly integrated into our daily lives. Just as routine dental and eye exams have become the norm, so will ongoing monitoring for our genetic predispositions and life-long exposures.

The fact that this book is written by a technology enthusiast and not a practicing physician is an indication that we are at the very beginning of a paradigm shift — the time of techno-savvy mavericks, start-up companies, and researcher-entrepreneurs. It is during this early developmental period that we have the greatest opportunities to influence the shape and direction of personalized medicine. These are truly revolutionary times. The future of medicine is being radically reshaped at this very moment — by you.

Index